CAMBRIDGE LIBRARY COLLECTION

Books of enduring scholarly value

Technology

The focus of this series is engineering, broadly construed. It covers technological innovation from a range of periods and cultures, but centres on the technological achievements of the industrial era in the West, particularly in the nineteenth century, as understood by their contemporaries. Infrastructure is one major focus, covering the building of railways and canals, bridges and tunnels, land drainage, the laying of submarine cables, and the construction of docks and lighthouses. Other key topics include developments in industrial and manufacturing fields such as mining technology, the production of iron and steel, the use of steam power, and chemical processes such as photography and textile dyes.

Samuel F.B. Morse

The American inventor Samuel Morse (1791–1872) spent decades fighting to be recognised for his key role in devising the electromagnetic telegraph. While he will always be remembered in the history of telecommunications, and for co-developing the code which bears his name, Morse started out as a painter and also involved himself in matters of politics over the course of his career. Published in 1914, this two-volume collection of personal papers was edited by his son, who provides helpful commentary throughout, illuminating the struggles and successes of a remarkable life. Volume 1 includes observations made in Europe while Morse studied painting. During the Napoleonic wars, he writes letters home describing the rising level of crime and social unrest in London, mentioning that he sleeps with a pistol. He is in London when Spencer Perceval is assassinated and later writes of meeting Turner, 'the best landscape painter living'.

Cambridge University Press has long been a pioneer in the reissuing of out-of-print titles from its own backlist, producing digital reprints of books that are still sought after by scholars and students but could not be reprinted economically using traditional technology. The Cambridge Library Collection extends this activity to a wider range of books which are still of importance to researchers and professionals, either for the source material they contain, or as landmarks in the history of their academic discipline.

Drawing from the world-renowned collections in the Cambridge University Library and other partner libraries, and guided by the advice of experts in each subject area, Cambridge University Press is using state-of-the-art scanning machines in its own Printing House to capture the content of each book selected for inclusion. The files are processed to give a consistently clear, crisp image, and the books finished to the high quality standard for which the Press is recognised around the world. The latest print-on-demand technology ensures that the books will remain available indefinitely, and that orders for single or multiple copies can quickly be supplied.

The Cambridge Library Collection brings back to life books of enduring scholarly value (including out-of-copyright works originally issued by other publishers) across a wide range of disciplines in the humanities and social sciences and in science and technology.

Samuel F.B. Morse

His Letters and Journals

VOLUME 1

SAMUEL FINLEY BREESE MORSE
EDITED BY EDWARD LIND MORSE

University Printing House, Cambridge, CB2 8BS, United Kingdom

Cambridge University Press is part of the University of Cambridge.
It furthers the University's mission by disseminating knowledge in the pursuit of education, learning and research at the highest international levels of excellence.

www.cambridge.org
Information on this title: www.cambridge.org/9781108074384

This edition first published 1914
This digitally printed version 2014

ISBN 978-1-108-07438-4 Paperback

SAMUEL F. B. MORSE

HIS LETTERS AND JOURNALS

IN TWO VOLUMES

VOLUME I

Saml. F. B. Morse

SAMUEL F. B. MORSE

HIS LETTERS AND JOURNALS

EDITED AND SUPPLEMENTED

BY HIS SON

EDWARD LIND MORSE

ILLUSTRATED
WITH REPRODUCTIONS OF HIS PAINTINGS
AND WITH NOTES AND DIAGRAMS
BEARING ON THE

INVENTION OF THE TELEGRAPH

VOLUME I

BOSTON AND NEW YORK
HOUGHTON MIFFLIN COMPANY
The Riverside Press Cambridge
1914

Published November 1914

TO MY WIFE

WHOSE LOVING INTEREST AND APT CRITICISM

HAVE BEEN TO ME OF GREAT VALUE

I DEDICATE THIS WORK

"It is the hour of fate,
And those who follow me reach every state
Mortals desire, and conquer every foe
Save death. But they who doubt or hesitate —
Condemned to failure, penury and woe —
Seek me in vain and uselessly implore.
I hear them not, and I return no more."

INGALLS, *Opportunity.*

PREFACE

ARTHUR CHRISTOPHER BENSON, in the introduction to his studies in biography entitled "The Leaves of the Tree," says: —

"But when it comes to dealing with men who have played upon the whole a noble part in life, whose vision has been clear and whose heart has been wide, who have not merely followed their own personal ambitions, but have really desired to leave the world better and happier than they found it, — in such cases, indiscriminate praise is not only foolish and untruthful, it is positively harmful and noxious. What one desires to see in the lives of others is some sort of transformation, some evidence of patient struggling with faults, some hint of failings triumphed over, some gain of generosity and endurance and courage. To slur over the faults and failings of the great is not only inartistic: it is also faint-hearted and unjust. It alienates sympathy. It substitutes unreal adoration for wholesome admiration; it afflicts the reader, conscious of frailty and struggle, with a sense of hopeless despair in the presence of anything so supremely high-minded and flawless."

The judgment of a son may, perhaps, be biased in favor of a beloved father; he may unconsciously "slur over the faults and failings," and lay emphasis only on the virtues. In selecting and putting together the letters, diaries, etc., of my father, Samuel F. B. Morse, I have tried to avoid that fault; my desire has been to present a true portrait of the man, with both lights and shadows duly emphasized; but I can say with perfect truth that I have found but little to deplore. He was human, he had his faults, and he made mistakes. While honestly differing from him on certain questions, I am yet convinced that, in all his beliefs, he was absolutely sincere, and the deeper I have delved into his correspondence, the more I have been impressed by the true nobility and greatness of the man.

His fame is now secure, but, like all great men, he made enemies who pursued him with their calumnies even after his death; and others, perfectly honest and sincere, have questioned his right to be called the inventor of the telegraph. I have tried to give credit where credit is due with regard to certain points in the invention, but I have also given the documentary evidence, which I am confident will prove that he never claimed more than was his right. For many years after his invention was a proved success, almost to the day of his death, he was compelled to fight for his rights; but he was a good fighter, a skilled controversialist, and he has won out in the end.

He was born and brought up in a deeply religious atmosphere, in a faith which seems to us of the present day as narrow; but, as will appear from his correspondence, he was perfectly sincere in his beliefs, and unfalteringly held himself to be an instrument divinely appointed to bestow a great blessing upon humanity.

It seems not to be generally known that he was an artist of great ability, that for more than half his life he devoted himself to painting, and that he is ranked with the best of our earlier painters.

In my selection of letters to be published I have tried

to place much emphasis on this phase of his career, a most interesting one. I have found so many letters, diaries, and sketch-books of those earlier years, never before published, that seemed to me of great human interest, that I have ventured to let a large number of these documents chronicle the history of Morse the artist.

Many of the letters here published have already appeared in Mr. S. Irenaeus Prime's biography of Morse, but others are now printed for the first time, and I have omitted many which Mr. Prime included. I must acknowledge my indebtedness to Mr. Prime for the possibility of filling in certain gaps in the correspondence; and for much interesting material not now otherwise obtainable.

Before the telegraph had demonstrated its practical utility, its inventor was subjected to ridicule most galling to a sensitive nature, and after it was a proved success he was vilified by the enemies he was obliged to make on account of his own probity, and by the unscrupulous men who tried to rob him of the fruits of his genius; but in this he was only paying the penalty of greatness, and, as the perspective of time enables us to render a more impartial verdict, his character will be found to emerge triumphant.

His versatility and abounding vitality were astounding. He would have been an eminent man in his day had he never invented the telegraph; but it is of absorbing interest, in following his career, to note how he was forced to give up one ambition after another, to suffer blow after blow which would have overwhelmed a man of less indomitable perseverance, until all his great energies were impelled into the one channel which ultimately led to undying fame.

In every great achievement in the history of progress one man must stand preëminent, one name must symbolize to future generations the thing accomplished, whether it be the founding of an empire, the discovery of a new world, or the invention of a new and useful art; and this one man must be so endowed by nature as to be capable of carrying to a successful issue the great enterprise, be it what it may. He must, in short, be a man of destiny. That he should call to his assistance other men, that he should legitimately make use of the labors of others, in no wise detracts from his claims to greatness. It is futile to say that without this one or that one the enterprise would have been a failure; that without his officers and his men the general could not have waged a successful campaign. We must, in every great accomplishment which has influenced the history of the world, search out the master mind to whom, under Heaven, the epoch-making result is due, and him must we crown with the laurel wreath.

Of nothing is this more true than of invention, for I venture to assert that no great invention has ever sprung Minerva-like from the brain of one man. It has been the culmination of the discoveries, the researches, yes, and the failures, of others, until the time was ripe and the destined man appeared. While due credit and all honor must be given to the other laborers in the field, the niche in the temple of fame must be reserved for the one man whose genius has combined all the known elements and added the connecting link to produce the great result.

As an invention the telegraph was truly epoch-making. It came at a time when steam navigation on land and water was yet in its infancy, and it is idle to speculate on the slow progress which this would have made had it not been for the assistance of the electric spark.

The science of electricity itself was but an academic curiosity, and it was not until the telegraph had demonstrated that this mysterious force could be harnessed to the use of man, that other men of genius arose to extend its usefulness in other directions; and this, in turn, stimulated invention in many other fields, and the end is not yet.

It has been necessary, in selecting letters, to omit many fully as interesting as those which have been included; barely to touch on subjects of research, or of political and religious discussion, which are worthy of being pursued further, and to omit some subjects entirely. Very probably another more experienced hand would have made a better selection, but my aim has been to give, through characteristic letters and contemporary opinions, an accurate portrait of the man, and a succinct history of his life and labors. If I have succeeded in throwing a new light on some points which are still the subject of discussion, if I have been able to call attention to any facts which until now have been overlooked or unknown, I shall be satisfied. If I have been compelled to use very plain language with regard to some of those who were his open or secret enemies, or who have been posthumously glorified by others, I have done so with regret.

Such as it is I send the book forth in the hope that it may add to the knowledge and appreciation of the character of one of the world's great men, and that it may, perhaps, be an inspiration to others who are striving, against great odds, to benefit their fellow men, or to those who are championing the cause of justice and truth.

Edward Lind Morse.

CONTENTS

CHAPTER I

APRIL 27, 1791 — SEPTEMBER 8, 1810

CHAPTER II

OCTOBER 31, 1810 — AUGUST 17, 1811

CHAPTER III

AUGUST 24, 1811 — DECEMBER 1, 1811

CHAPTER IV

JANUARY 18, 1812 — AUGUST 6, 1812

CHAPTER V

SEPTEMBER 20, 1812 — JUNE 13, 1813

CHAPTER VI

JULY 10, 1813 — APRIL 6, 1814

CHAPTER XIII

JANUARY 4, 1825 — NOVEMBER 18, 1825

CHAPTER XIV

JANUARY 1, 1826 — DECEMBER 5, 1829

CHAPTER XV

DECEMBER 6, 1829 — FEBRUARY 6, 1830

CHAPTER XVI

FEBRUARY 6, 1830—JUNE 15, 1830

CHAPTER XVII

JUNE 17, 1830—FEBRUARY 2, 1831

CHAPTER XVIII

FEBRUARY 10, 1831—SEPTEMBER 12, 1831

CHAPTER XIX

SEPTEMBER 18, 1831 — SEPTEMBER 21, 1832

CHAPTER XX

ILLUSTRATIONS

SAMUEL F. B. MORSE
HIS LETTERS AND JOURNALS

CHAPTER I

APRIL 27, 1791 — SEPTEMBER 8, 1810

Birth of S. F. B. Morse. — His parents. — Letters of Dr. Belknap and Rev. Mr. Wells. — Phillips, Andover. — First letter. — Letter from his father. — Religious letter from Morse to his brothers. — Letters from the mother to her sons. — Morse enters Yale. — His journey there. — Difficulty in keeping up with his class. — Letter of warning from his mother. — Letters of Jedediah Morse to Bishop of London and Lindley Murray. — Morse becomes more studious. — Bill of expenses. — Longing to travel and interest in electricity. — Philadelphia and New York. — Graduates from college. — Wishes to accompany Allston to England, but submits to parents' desires.

SAMUEL FINLEY BREESE MORSE was born in Charlestown, Massachusetts, on the 27th day of April, A.D. 1791. He came of good Puritan stock, his father, Jedediah Morse, being a militant clergyman of the Congregational Church, a fighter for orthodoxy at a time when Unitarianism was beginning to undermine the foundations of the old, austere, childlike faith.

These battles of the churches seem far away to us of the twentieth century, but they were very real to the warriors of those days, and, while many of the tenets of their faith may seem narrow to us, they were gospel to the godly of that time, and reverence, obedience, filial piety, and courtesy were the rule and not the exception that they are to-day.

Jedediah Morse was a man of note in his day, known and respected at home and abroad; the friend of General Washington and other founders of the Republic; the author of the first American Geography and Gazetteer. His wife, Elizabeth Ann Breese, granddaughter of Samuel Finley, president of Princeton College, was a woman of great strength and yet sweetness of character; adored by her family and friends, a veritable mother in Israel.

Into this serene home atmosphere came young Finley Morse, the eldest of eleven children, only three of whom survived their infancy. The other two were Sidney Edwards and Richard Carey, both eminent men in their day.

Dr. Belknap, of Boston, in a letter to a friend in New York says:—

"Congratulate the Monmouth Judge [Mr. Breese] on the birth of a grandson. . . . As to the child, I saw him asleep, so can say nothing of his eye or his genius peeping through it. He may have the sagacity of a Jewish rabbi, or the profundity of a Calvin, or the sublimity of a Homer for aught I know. But time will show forth all things."

This sounds almost prophetic in the light of future days.

The following letter from the Reverend Mr. Wells is quaint and characteristic of the times: —

My dear little Boy, — As a small testimony of my respect and obligation to your excellent Parents and of my love to you, I send you with this six (6) English Guineas. They are pretty playthings enough, and in the Country I came from many people are fond of them. Your Papa will let you look at them and shew them to Edward, and then he will take care of them,

HOUSE IN WHICH MORSE WAS BORN, IN CHARLESTOWN, MASS.

and, by the time you grow up to be a Man, they will under Papa's wise management increase to double their present number. With wishing you may never be in want of such playthings and yet never too fond of them, I remain your affectionate friend,

WM. WELLS.

MEDFORD, July 2, 1793.

Young Morse was sent away early to boarding-school, as was the custom at that time. He was taken by his father to Phillips Academy at Andover, and I believe he ran away once, being overcome by homesickness before he made up his mind to remain and study hard.

The following letter is the first one written by him of which I have any knowledge: —

ANDOVER, 2d August, 1799.

DEAR PAPA, — I hope you are well I will thank you if you will Send me up Some quils Give my love to mama and NANCY and my little brothers pleas to kis them for me and send me up Some very good paper to write to you

I have as many blackberries as I want I go and pick them myself.

SAMUEL FINLEY BREESE MORSE
YOUR SON
1799.

This from his father is characteristic of many written to him and to his brothers while they were at school and college: —

CHARLESTOWN, February 21, 1801.

MY DEAR SON, — You do not write me as often as you ought. In your next you must assign some reason for this neglect. Possibly I have not received all your letters. Nothing will improve you so much in epistolary writing as practice. Take great pains with your letters. Avoid vulgar phrases. Study to have your ideas pertinent and correct and clothe them in an easy and grammatical dress. Pay attention to your spelling, pointing, the use of capitals, and to your handwriting. After a little practice these things will become natural and you will thus acquire a habit of writing correctly and well.

General Washington was a remarkable instance of what I have now recommended to you. His letters are a perfect model for epistolary writers. They are written with great uniformity in respect to the handwriting and disposition of the several parts of the letter. I will show you some of his letters when I have the pleasure of seeing you next vacation, and when I shall expect to find you much improved.

Your natural disposition, my dear son, renders it proper for me earnestly to recommend to you to *attend to one thing at a time.* It is impossible that you can do two things well at the same time, and I would, therefore, never have you attempt it. Never undertake to do what ought not to be done, and then, whatever you undertake, endeavor to do it in the best manner.

It is said of De Witt, a celebrated statesman in Holland, who was torn to pieces in the year 1672, that he did the whole business of the republic and yet had time left to go to assemblies in the evening and sup in company. Being asked how he could possibly find time to go through so much business and yet amuse himself in the evenings as he did, he answered there was nothing so easy, for that it was only doing one thing at a time, and never putting off anything till to-morrow that could be done to-day. This steady and undissipated attention to one object is a sure mark of a superior genius, as hurry, bustle, and agitation are the never-failing symptoms of a weak and frivolous mind.

I expect you will read this letter over several times that you may retain its contents in your memory, and give me your own opinion on the advice I have given you. If you improve this well, I shall be encouraged to give you more as you may need it.

Your affectionate parent,

J. Morse.

This was written to a boy ten years old. I wonder if he was really able to assimilate it.

I shall pass rapidly over the next few years, for, while there are many letters which make interesting reading, there are so many more of the later years of greater historical value that I must not yield to the temptation to linger.

The three brothers were all sent to Phillips Academy to prepare for Yale, from which college their father was also graduated.

The following letter from Finley to his brothers was written while he was temporarily at home, and shows the deep religious bent of his mind which he kept through life: —

Charlestown, March 15, 1805.

My dear Brothers, — I now write you again to inform you that mama had a baby, but it was born dead and has just been buried. Now you have three brothers and three sisters in heaven and I hope you and I will meet them there at our death. It is uncertain when we shall die, but we ought to be prepared for it, and I hope you and I shall.

I read a question in Davie's "Sermons" the last Sunday which was this: — Suppose a bird should take one dust of this earth and carry it away once in a thousand years, and you was to take your choice either to be miserable in that time and happy hereafter, or happy in that time and miserable hereafter, which would you choose? Write me an answer to this in your next letter. . . .

I enclose you a little book called the "Christian Pilgrim." It is for both of you.

We are all tolerable well except mama, though she is more comfortable now than she was. We all send a great deal of love to you. I must now bid you adieu.

I remain your affectionate brother,

S. F. B. Morse.

I am tempted to include the following extracts from letters of the good mother of the three boys as characteristic of the times and people: —

Charlestown, June 28, 1805.

My dear Son, — We have the pleasure of a letter from you which has gratified us very much. It is the only intelligence we have had from you since Mr. Brown left you. I began to think that something was the matter with respect to your health that occasioned your long silence. . . . We are very desirous, my son, that you should excel in everything that will make you truly happy and useful to your fellow men. In particular by no means neglect your duty to your Heavenly Father.

Remember, what has been said with great truth, that he can never be faithful to others who is not so to his God and his conscience. I wish you constantly to keep in mind the first question and answer in that excellent form of sound words, the Assembly Catechism, viz: — "What is the chief end of Man?" The answer you will readily recollect is "To Glorify God and enjoy Him forever."

Let it be evident, my dear son, that this be your chief aim in all that you do, and may you be so happy as to enjoy Him forever is the sincere prayer of your affectionate parent. . . .

The Fourth of July is to be celebrated here with a good deal of parade both by Federalists and Jacobins. The former are to meet in our meeting-house. there to hear an oration which is to be delivered by Mr. Aaron Putnam, a prayer by your papa also. And on the hill close by the monument [Bunker Hill] a standard is to be presented to a new company called the Warren Phalanx, all Federalists, by Dr. Putnam who is the president of the day, and all the gentlemen are to dine at Seton's Hall, otherwise called Massachusetts Hall, and the ladies are to take tea at the same place. The Jacobins are to have an oration at the Baptist meeting-house from Mr. Gleson. I know nothing more about them. The boys are forming themselves into companies also; they have two or three companies and drums which at some times are enough to craze one. I can't help thinking when I see them how glad I am that my sons are better employed at Andover than beating the streets or drums; that they are laying in a good store of useful knowledge against the time to come, while these poor boys, many of them, at least, are learning what they will be glad by and by to unlearn.

July 30, 1805.

MY DEAR SONS, — Have you heard of the death of young Willard at Cambridge, the late President Willard's son? He died of a violent fever occasioned by going into water when he was very hot in the middle of the day. He also pumped a great deal of cold water on his head. Let this be a warning to you all not to be guilty of the like indiscretion which may cost you your life. Dreadful, indeed, would this be to all of us. I wish you would not go into water oftener than once a week, and then either early in the morning or late in the afternoon, and not go in when hot nor stay long in the water. Remember these cautions of your mama and obey them strictly.

A young lady twenty years old died in Boston yesterday very suddenly. She eat her dinner perfectly well and was dead in five minutes after. Her name was Ann Hinkley. You see, my dear boys, the great uncertainty of life and, of course, the importance of being always prepared for *death*, even a *sudden death*, as we know not what an hour may bring forth. This we are sensible of, we cannot be *too soon* or *too well* prepared for that all-important moment, as this is what we are sent into this world for. The main business of life is to prepare for death. Let us not, then, put off these most important concerns to an uncertain to-morrow, but let us in earnest attend to the concerns of our precious, never-dying souls while we feel ourselves alive.

In October, 1805, Finley Morse went to New Haven to enter college, and the next letter describes the journey from Charlestown, and it was, indeed, a journey in those days.

NEW HAVEN, October 22, 1805.

MY DEAR PARENTS, — I arrived here yesterday safe and well. The first day I rode as far as Williams' Tavern, and put up there for the night. The next day I rode as far as Dwight's Tavern in Western, and in the morning, it being rainy, Mr. Backus did not set out to ride till late, and, the stage coming to the door, Mr. B. thought it a good opportunity to send me to Hartford, which he did, and I arrived at Hartford that night and lodged at Ripley's inn opposite the State House. He treated me very kindly, indeed, wholly on account of my being your son. I was treated more like his own son than a stranger, for which I shall and ought to be very much obliged to him. The next morning I hired a horse and chaise of him to carry me to Weathersfield and arrived at Mr. Marsh's, who was very glad to see me and begged me to stay till S. Barrell went, which was the next Monday, for his mother would not let him go so soon, she was so glad to see him. I was sorry to trouble them so much, but, as they desired it, and, as Samuel B. was not to go till then, I agreed to stay and hope you will not disapprove it, and am sorry I could not write you sooner to relieve your minds from your anxiety on my account, and am sorry for giving my good parents so much trouble and expense. You expend and have expended a great deal more money upon me than I deserve, and granted me a great many of my requests, and I am sure I can certainly grant you one, that of being *economical*, which I shall certainly be and not get money to buy trifling things. I begin to think *money* of some importance and too great value to be thrown away.

Yesterday morning about ten o'clock I set out for New Haven with S. Barrell and arrived well a little before dark. I went directly to Dr. Dwight's, which I easily found, and delivered the letter to him, drank tea at his house, and then Mr. Sereno Dwight carried me to Mr. Davis's who had agreed to take me. While I was at Dr. Dwight's there was a woman there whom the Dr. recommended to Sam. B. and me to have our mending done, and Mrs. Davis or a washerwoman across the way will do my washing, so I am very agreeably situated. I also gave the letter to Mr. Beers and he has agreed to let me have what you desired. I have got Homer's Iliad in two volumes, with Latin translation of him, for $3.25. I need no other books at present.

S. Barrell has a room in the north college and, as he says, a very agreeable chum.

Next spring I hope you will come on and fix matters. I long to get into the college, for it appears to me now as though I was not a member of college but fitting for college. I hope next spring will soon come.

My whole journey from Charlestown here cost me £2 16*s*., and 4*d*., a great deal more than either you or I had calculated on. I am sorry to be of so much trouble to you and the cause of so much anxiety in you and especially in mama. I wish you to give my very affectionate love to my dear brothers, and tell them they must write me and not be homesick, but consider that I am farther from home than they are, 136 miles from home. I remain

Your ever affectionate son,
S. F. B. Morse.

It would seem, from other letters which follow, that he had difficulty in keeping up with his class, and that he eventually dropped a class, for he did not graduate until 1810. He also seems to have been rooming outside of college and to have been eager to go in.

It is curious, in the light of future events, to note that young Morse's parents were fearful lest his volatile nature and lack of steadfastness of purpose should mar his future career. His dominating characteristic in later life was a bulldog tenacity, which led him to stick to one idea through discouragements and disappointments which would have overwhelmed a weaker nature.

The following extracts are from a long letter from his mother dated November 23, 1805: —

"I am fearful, my son, that you think a great deal more of your amusements than your studies, and there lies the difficulty, and the same difficulty would exist were you in college.

"You have filled your letter with requests to go into college and an account of a gunning party, both of which have given us pain. I am truly sorry that you appear so unsteady as by *your own account* you are. . . .

"You mention in the letter you wrote first that, if you went into college, you and your chum would want brandy and wine and segars in your room. Pray is that the custom among the students? We think it a very improper one, indeed, and hope the government of college will not permit it. There is no propriety at all in such young boys as you having anything to do with anything of the kind, and your papa and myself positively prohibit you the use of these things till we think them more necessary than we do at present. . . .

"You will remember that you have promised in your first letter to be an economist. In your last letter you seem to have forgotten all about it. Pray, what do your gunning parties cost you for powder and shot? I beg you to consider and not go driving on from one foolish whim to another till you provoke us to withdraw from you the means of gratifying you in anything that may be even less objectionable than gunning."

These exhortations seem to have had, temporarily, at least, the desired effect, for in a letter to his parents dated December 18, 1805, young Morse says: "I shall not go out to gun any more, for I know it makes you anxious about me."

The letters of the parents to the son are full of pious exhortations, and good advice, and reproaches to the boy for not writing oftener and more at length, and for not answering every question asked by the parents. It is comforting to the present-day parent to learn that human nature was much the same in those pious days of old, differing only in degree, and that there is hope for the most wayward son and careless correspondent.

The following letters from the elder Morse I shall include as being of rather more than ordinary interest, and as showing the breadth of his activity.

CHARLESTOWN, December 23, 1806.

TO THE BISHOP OF LONDON,

REV'D AND RESPECTED SIR, — I presume that it might be agreeable to you to know the precise state of the property which originally belonged to the Protestant Episcopal Church in Virginia.

I have with some pains obtained the law of that State respecting this singular business.

I find that it destroys *the establishment* and asserts that "all property belonging to the said (Protestant Episcopal) Church devolved on the good people of this Commonwealth (i.e., Virginia) on the dissolution of the British Government here, in the same degree in which the right and interest of the said Church was therein derived from them," and authorizes the overseers of the poor of any county "in which any glebe land is vacant, or shall become so by the death or removal of any incumbent, to sell all such land and appurtenances and every other species of property incident thereto to the highest bidder" — "Provided that nothing herein contained shall authorize an appropriation to *any religious purpose whatever*."

I make no comments on the above. I believe no other State in the Union has, in this respect, imitated the example of Virginia.

I take the liberty to send you a few small tracts for your acceptance in token of my high respect for your character and services.

Believe me, sir, unfeignedly,

Your obedient servant,

J. MORSE.

December 26, 1806.

LINDLEY MURRAY ESQ.,

DEAR SIR, — Your polite note and the valuable books accompanying it, forwarded by our friend Perkins, of New York, have been duly and gratefully received.

You will perceive, by the number of the "Panoplist" enclosed, that we are strangers neither to your works nor your character. It has given me much pleasure as an American to make both more extensively known among my countrymen.

I have purchased several hundred of your spelling books for a charitable society to which I belong, and they have been dispersed in the new settlements in our country, where I hope they will do immediate good, besides creating a desire and demand for more. It will ever give me pleasure to hear from you when convenient. Letters left at Mr. Taylor's will find me.

I herewith send you two or three pamphlets and a copy of the last edition of my "American Gazetteer" which I pray you to accept as a small token of the high respect and esteem with which I am

Your friend,

J. MORSE.

Young Morse now settled down to serious work as the following extracts will show, which I set down without further comment, passing rapidly over the next few years. He was, however, not entirely absorbed in his books but still longed for the pleasures of the chase: —

"May 13, 1807. Just now I asked Mr. Twining to let me go a-gunning for this afternoon. He told me you had expressly forbidden it and he therefore could not. Now I should wish to go once in a while, for I always intend to be careful. I have no amusement now in the vacation, and it would gratify me very much if you would consent to let me go once in a while. I suppose you would tell me that my books ought to be my amusement. I cannot study all the time and I need some exercise. If I walk, that is no amusement, and if I wish to play ball or anything else, I have no one to play with. Please to write me an answer as soon as possible."

June 7, 1807.

My dear Parents, — I hope you will excuse my not writing you sooner when I inform you that my time is entirely taken up with my studies.

In the morning I must rise at five o'clock to attend prayers and, immediately after, recitation; then I must breakfast and begin to study from eight o'clock till eleven; then recite my forenoon's lesson which takes me an hour.

At twelve I must study French till one, which is dinner-time. Directly after dinner I must recite French to Monsieur Value till two o'clock, then begin to study my afternoon lesson and recite it at five. Immediately after recitation I must study another French lesson to recite at seven in the evening; come home at nine o'clock and study my morning's lesson until ten, eleven, and sometimes twelve o'clock, and by that time I am prepared to sleep. . . . You see now I have enough to do, my hands as full as can be, not five minutes' time to take recreation. I am determined to study and, thus far, have not missed a single word. The students call me by the nickname of "Geography."

"*June 18, 1807.* Last week I went to Mr. Beers and saw a set of Montaigne's 'Essays' in French in eight volumes, duodecimo, handsomely bound in calf and gilt, for two dollars. The reason they are so cheap is because they are wicked and bad books for me or anybody else to read. I got them because they were cheap, and have exchanged them for a handsome English edition of 'Gil Blas'; price, $4.50."

In the fall of 1807 Finley Morse returned to college accompanied by his next younger brother, Sidney Edwards. In a letter of March 6, 1808, he says: "Edwards and myself are very well and I believe we are doing well, but you will learn more of that from our instructors."

In this same letter he says: —

"I find it impossible to live in college without spending money. At one time a letter is to be paid for, then comes up a great tax from the class or society, which keeps me constantly running after money. When I have money in my hand I feel as though I had stolen it, and it is with the greatest pain that I part with it. I think every minute I shall receive a letter from home blaming me for not being more economical, and thus I am kept in distress all the time.

"The amount of my expenses for the last term was fifteen dollars, expended in the following manner: —

	Dols. Cts.	
"Postage	$2.05	
Oil	.50	
Taxes, fines, etc.	3.00	
Oysters	.50	
Washbowl	.37½	
Skillet	.33	
Axe $1.33 Catalogues .12	1.45	
Powder and shot	1.12½	
Cakes, etc. etc. etc.	1.75	
Wine, Thanks. day	.20	
Toll on bridge	.15	
Grinding axe	.08	
Museum	.25	
Poor man	.14	
Carriage for trunk	1.00	
Pitcher	.41	14.75½
Sharpening skates	.37½	Paid for
Circ. Library	.25	cutting wood .25
Post papers	.57	
Lent never to be returned	.25	
	$14.75½	$15.00½

"In my expenses I do not include my wood, tuition bills, board or washing bills."

How characteristic of all boys of all times the "etc., etc., etc.," tacked on to the "cakes" item, and how many boys of the present day would bewail the extravagance of fifteen dollars spent in one term on extras? In a postscript in this same letter he says: "The students are very fond of raising balloons at present. I will (with your leave) when I return home make one. They are pleasant sights."

College terms were very different in those days from what they are at present, for September 5 finds the boys still in New Haven, and Finley says, "There is but three and a half weeks to Commencement."

In this same letter he gives utterance to these filial

sentiments: "I now make those only my companions who are the most religious and moral, and I hope sincerely that it will have a good effect in changing that thoughtless disposition which has ever been a striking trait in my character. As I grow older, I begin to think better of what you have always told me when I was small. I begin to know by experience that man is born to trouble, and that temptations to do evil are as countless as the stars, but I hope I shall be enabled to shun them."

This is from a letter of January 9, 1809: —

"I have been reading the first volume of Professor Silliman's 'Journal' which he kept during his passage to and residence in Europe. I am very much pleased with it. I long for the time when I shall be able to travel with improvement to myself and society, and hope it will be in your power to assist me.

"I have a very ardent desire of travelling, but I consider that an education is indispensable to me and I mean to apply myself with all diligence for that purpose. *Diligentia vincit omnia* is my maxim and I shall endeavor to follow it. . . . I shall be employed in the vacation in the Philosophical Chamber with Mr. Dwight, who is going to perform a number of experiments in *Electricity*."

It is, of course, only a curious coincidence that these two sentences should have occurred in the same letter, but it was when travelling, many years afterwards, that the first idea of the electric telegraph found lodgment in his brain, and this certainly resulted in improvement to himself and society.

In February, 1809, he writes: "My studies are at

present Optics in Philosophy, Dialling, Homer, beside disputing, composing, attending lectures etc. etc., all which I find very interesting and especially Mr. Day's lectures who is now lecturing on *Electricity*."

Young Morse's thoughts seem to have been gradually focusing on the two subjects to which he afterwards devoted his life, for in a letter of March 8, 1809, he says: "Mr. Day's lectures are very interesting. They are upon Electricity. He has given us some very fine experiments. The whole class taking hold of hands formed the circuit of communication and we all received the shock apparently at the same moment. I never took an electric shock before. It felt as if some person had struck me a slight blow across the arms. . . . I think with pleasure that two thirds of this term only remain. As soon as that is passed away, I hope I shall again see home. I really long to see Charlestown again; I have almost forgotten how it looks. I have some thoughts of taking a view of Boston from Bunker's Hill when I go home again. It will be some pleasure to me to have some picture of my native place to look upon when I am from home."

And in August, 1809, he writes to his parents: "I employ all my leisure time in painting. I have a great number of persons engaged already to be drawn on ivory, no less than seven. They obtain the ivories for themselves. I have taken Professor Kingsley's profile for him. It is a good likeness of him and he is pleased with it. I think I shall take his likeness on ivory and present it to him as my present at the end of the year. . . . I have finished Miss Leffingwell's miniature. It is a good likeness and she is very much pleased with it."

New Haven, May 29, 1810.

My dear Parents, — I arrived in this place on Sabbath evening by packet from New York. I left Philadelphia on Thursday morning at eight o'clock and arrived in New York on Friday at ten. . . .

I stayed in New York but one night. I found it quite insipid after seeing Philadelphia. [The character of the two cities seems to have changed a trifle in a hundred years, for, with all her faults, no one could nowadays accuse New York of being insipid.] I went on board the packet on Saturday at twelve o'clock and arrived, as I before stated, on Sabbath evening. We had, on the whole, a very good set of passengers from New York to this place. On Sunday we had two sermons read to us by one of them, Dr. Hawley, of this place, and in the evening we sang five psalms, and during the whole of the exercises the passengers conducted themselves with perfect decorum, although one of the sermons was one hour in length. . . .

June 25, 1810.

My dear Parents, — I received yours of the 23d this day and receive with humility your reproof. I am extremely sorry it should have occasioned so many disagreeable feelings. I felt it my duty to tell you of my debts, and, indeed, I could not feel easy without. The amount of my buttery bill is forty-two or forty-three dollars.

Mr. Nettleton is butler and is willing I should take his likeness as part pay. I shall take it on ivory, and he has engaged to allow me seven dollars for it. My price is five dollars for a miniature on ivory, and I have engaged three or four at that price. My price for profiles is one dollar, and everybody is ready to engage me at that price. . . . Though I have been much to blame in the present case, yet I think it but just that Mr. Twining should bear his part.

I had begun with a determination to pay for everything as I got it, but was stopped in this in the very beginning, for, in going to Mr. T. to get money, I have five times out of six found him absent, sometimes for the whole day, sometimes for a week or two weeks, and once he was absent six weeks and made no sort of provision for us. Mrs. T. is never trusted with money for us. Now in such case I am obliged by necessity to get a thing charged, and I have found by sad experience that a bill increases faster than I had in the least imagined. . . .

"*July 22, 1810.* I am now released from college and am attending to painting. All my class were accepted as candidates for degrees. Edwards is admitted a member of ΦBK Society, and is appointed as monitor to the next Freshman Class. Richard is chosen as one of the speakers the evening before Commencement.

"Edwards and Richard are both of them very steady and good scholars, and are much esteemed by the authority of college as well as their fellow students.

"As to my choice of a profession, I still think that I was made for a painter, and I would be obliged to you to make such arrangement with Mr. Allston for my studying with him as you shall think expedient. I should desire to study with him during the winter, and, as he expects to return to England in the spring, I should admire to be able to go with him."

In answer to this letter his father wrote: —

CHARLESTOWN, July 26, 1810.

DEAR FINLEY, — I received your letter of the 22d to-day by mail.

On the subject of your future pursuits we will converse when I see you and when you get home. It will be best for you to form no plans. Your mama and I have been thinking and planning for you. I shall disclose to you our plan when I see you. Till then suspend your mind.

It gives us great pleasure to have you speak so well of your brothers. Others do the same and we hear well of you also. It is a great comfort to us that our sons are all likely to do so well and are in good reputation among their acquaintances. Could we have reason to believe you were all pious and had chosen the "good part," our joy concerning you all would be full. I hope the Lord in due time will grant us this pleasure.

"Seek the Lord," my dear son, "while he may be found."

Your affectionate father,
J. MORSE.

September 8, 1810.

DEAR MAMA, — Papa arrived here safely this evening and I need not tell you we were glad to see him. He has mentioned to me the plan which he proposed for my future business in life, and I am pleased with it, for I was determined beforehand to conform to his and your will in everything, and, when I come home, I

ELIZABETH ANN MORSE AND SIDNEY E. MORSE

REV. JEDEDIAH MORSE AND S. F. B. MORSE

From portraits by a Mr. Sargent, who also painted portraits of the Washington family

shall endeavor to make amends for the trouble and anxiety which you have been at on my account, by assisting papa in his labors and pursuing with ardor my own business. . . .

I have been extremely low-spirited for some days past, and it still continues. I hope it will wear off by Commencement Day. . . .

I am so low in spirits that I could almost cry.

It was no wonder that he was down-hearted, for he was ambitious and longed to carve out a great career for himself, while his good parents were conservative and wished him to become independent as soon as possible. Their plan was to apprentice him to a bookseller, and he dutifully conformed to their wishes for a time, but his ambition could not be curbed, and it was not long before he broke away.

CHAPTER II

OCTOBER 31, 1810—AUGUST 17, 1811

Enters bookshop as clerk. — Devotes leisure to painting. — Leaves shop. — Letter to his brothers on appointments at Yale. — Letters from Joseph P. Rossiter. — Morse's first love affair. — Paints "Landing of the Pilgrims." — Prepares to sail with Allstons for England. — Letters of introduction from his father. — Disagreeable stage-ride to New York. — Sails on the Lydia. — Prosperous voyage. — Liverpool. — Trip to London. — Observations on people and customs. — Frequently cheated. — Critical time in England. — Dr. Lettsom. — Sheridan's verse. — Longing for a telegraph. — A ghost.

After his graduation from Yale College in the fall of 1810, Finley Morse returned to his home in Charlestown, Mass., and cheerfully submitted himself to his parents' wishes by entering the bookshop of a certain Mr. Mallory.

He writes under date of October 31, 1810, to his brothers who are still at college: "I am in an excellent situation and on excellent terms. I have four hundred dollars per year, but this you must not mention out. I have the choice of my hours; they are from nine till one-half past twelve, and from three till sunset."

But he still clings to the idea of becoming a painter, for he adds: "My evenings I employ in painting. I have every convenience; the room over the kitchen is fitted up for me; I have a fire there every evening, and can spend it alone or otherwise as I please. I have bought me one of the new patent lamps, those with glass chimneys, which gives an excellent light. It cost me about six dollars. Send on as soon as possible anything and everything which pertains to my painting apparatus."

The following letter was written at some time in 1810 or 1811. It was addressed to Mr. Sereno E. Dwight: —

"Mr. Mallory a few days since handed me a letter from you requesting me, if possible, to sketch a likeness of young Mr. Daggett. Accordingly I have made the attempt and take the present opportunity of forwarding you the results. The task was hard but pleasurable. It is one of the most difficult undertakings to endeavor to take a portrait from recollection of one whose countenance has not been examined particularly for the purpose. When I made the first attempt, not a single feature could I recall distinctly to my memory and I almost despaired of a likeness, but the thought of lessening the affliction of such a distressed family determined me to attempt it a second time. The result is on the ivory. I then showed it to my brothers, to Mr. Evarts, to Mr. Hillhouse, to Mr. Mallory, and to Mr. Read, all of whom had not the least suspicion of anything of the kind, and they have severally and separately pronounced it a likeness of young Mr. Daggett. This encouraged me, and I made the two other sketches which are thought likewise to be resemblances of him.

"If these or any one of them can be recognized by the afflicted family as a resemblance of him they have lost, it will be an ample compensation to me to think that I have in any degree been the means of alleviating their suffering. . . ."

On December 8, 1810, he writes to his brother: "I have almost completed my landscape. It is 'proper handsome,' so they say, and they want to make me believe it is so, but I shan't yet awhile."

This shows the right frame of mind for an artist,

and yet, like most youthful painters, he attempted more than his proficiency warranted, for in this same letter he adds: "I am going to begin, as soon as I have finished it [the landscape], a piece, the subject of which will be 'Marius on the Ruins of Carthage.'"

On December 28, 1810, he writes: "I shall leave Mr. Mallory's next week and study painting exclusively till summer."

He had at last burst his bonds, and his wise parents, seeing that his heart was only in his painting, decided to throw no further obstacles in his way, but, at the cost of much self-sacrifice on their part, to further in every way his ambition.

January 15, 1811.

My dear Brothers, — We have just received Richard's letter of the 8th inst., and I can have a pretty correct idea of your feelings at the beginning of a vacation. You must not be melancholy and hang yourself. If you do you will have a terrible scolding when you get home again. As for Richard's getting an appointment so low, if I was in his situation, I should not trouble myself one fig concerning *appointments*. They cost more than they are worth. I shall not esteem him the less for not getting a higher, and not more than one millionth part of the world knows what an appointment is. You will both of you have a different opinion of appointments after you have been out of college a short time. I had rather be Richard with a dialogue than Sanford with a dispute. If appointments at college decided your fate forever, you might possibly groan and wail. But then consider where poor I should come. [He got no appointment whatever.] Think of this,

Richard, and *don't* hang *yourself*. [It may, perhaps, be well to explain that "appointments" were given at Yale to those who excelled in scholarship. "Philosophical Oration" was the highest, then came "High Oration," "Oration," etc., etc.] I have left Mr. Mallory's store and am helping papa in the Geography. Shall remain at home till the latter part of next summer and then shall go to London with Mr. Allston.

The following extracts from two letters of a college friend I have introduced as throwing some light on Morse's character at that time and also as curious examples of the epistolary style of those days: —

NEW HAVEN, February 5, 1811.

DEAR FINLEY, — Yours of the 6th ult. I received, together with the books enclosed, which I delivered personally according to your request.

Did I not know the nature of your disorder and the state of your *gizzard*, I should really be surprised at the commencement, and, indeed, the whole tenor of your letter, but as it is I can excuse and feel for you.

Had I commenced a letter with the French *Hélas! hélas!* it would have been no more than might reasonably have been expected considering the desolate situation of New Haven and the gloomy prospects before me. But for you, who are in the very vortex of fashionable life and surrounded by the amusements and bustle of the metropolis of New England, for you to exclaim, "How lonely I am!" is unpardonable, or at most admits of but one excuse, to wit, that you can plead the feelings of the youth who exclaimed, "Gods annihilate both time and space and make two lovers happy!"

You suppose I am so much taken up with the ladies and other good things in New Haven that I have not time to think of one of my old friends. Alas! Morse, there are no ladies or anything else to occupy my attention. They are all gone and we have no amusements. Even old Value has deserted us, whose music, though an assemblage of "unharmonious sounds," is infinitely preferable to the harsh grating thunder of his brother.

New Haven is, indeed, this winter a dreary place. I wrote you about a month since and did then what you wish me now to do, — I mentioned all that is worth mentioning, which, by the way, is very little, about New Haven and its inhabitants.

Since then I have been to New York and saw the Miss Radcliffs, and, in passing through Stamford, the Miss Davenports. The mention of the name of Davenport would at one time have excited in your breast emotions unutterable, but now, though Ann is as lovely as ever, your heart requires the influence of another Hart to quicken its pulsations. . . . Last but not least comes the all-conquering, the angelic queen of Harts. I have not seen her since she left New Haven, but have heard from her sister Eliza that she is in good health and is going in April to New York with Mrs. Jarvis (her sister) to spend the summer and perhaps a longer time, where she will probably break many a proud heart and bend many a stubborn knee. I fear, Morse, unless you have her firmly in your toils, I fear she may not be able to withstand every attack, for New York abounds with elegant and accomplished young men.

You mention that you have again changed your mind as to the business which you intend to pursue. I really thought that the plan of becoming a bookseller would be permanent because sanctioned by parental authority, but I am now convinced that your mind is so much bent upon painting that you will do nothing else effectually. It is indeed a noble art and if pursued effectually leads to the highest eminence, for painters rank with poets, and to be placed in the scale with Milton and Homer is an honor that few of mortal mould attain unto. . . . I wish, Finley, that you would paint me a handsome piece for a keepsake as you are going to Europe and may not be back in a hurry. Present my respects to Mr. Hillhouse. His father's family are well. Adieu.

Your affectionate friend,
Jos. P. Rossiter.

From this letter and from others we learn that young Morse's youthful affections were fixed on a certain charming Miss Jannette Hart, but, alas! he proved a faithless lover, for his friend Rossiter thus reproves him in a letter of May 8, 1811: —

"Oh! most amazing change! Can it be possible? Oh! Love, and all ye cordial powers of passion, forbid it! Still, still the dreadful words glare on my sight. Alas! alas! and is it, then, a fact? If so 't is pitiful, 't is wondrous pitiful. Cupid, tear off your bandage, new string your bow and tip your arrows with harder adamant. Oh! shame upon you, only hear the words of your exultant votarist — 'Even Love, which according to the proverb conquers all things, when put in competition with painting, must yield the palm and be a willing captive.' Oh! fie, fie, good master Cupid, you shoot but poorly if a victim so often wounded can talk in terms like these.

"Poor luckless Jannette! the epithets 'divine' and 'heavenly' which have so often been applied to thee are now transferred to miserable daubings with oil and clay. Dame Nature, your triumph has been short. Poor foolish beldam, you thought, indeed, when you had formed your masterpiece and named her Jannette, that unqualified admiration would be extorted from the lips of prejudice itself, and that, at least, till age had worn off the first dazzling lustre from your favorite, your sway would have been unlimited and your exultation immeasurable. My good old Dame, hear for your comfort what a foolish, fickle youth has dared to say of your darling Jannette, and that while she is yet in the first blush and bloom of virgin loveliness — '*next* to painting I love Jannette the best.' Insufferable blasphemy! Hear, O Heavens, and be amazed! Tremble, O Earth, and be horribly afraid!"

In spite of this impassioned arraignment, Morse devoted himself exclusively to his art for the next few years, and we have only occasional references in the letters that follow to his first serious love affair.

We also hear nothing further of "Marius on the Ruins of Carthage"; but in February, 1811, he writes to his brothers: "I am painting my large piece, the landing of our forefathers at Plymouth. Perhaps I shall have it finished by the time you come home in the spring. My landscape I finished sometime since, and it is framed and hung up in the front parlor."

At last in July, 1811, the great ambition of the young man was about to be realized and he prepared to set sail for England with his friend and master, Washington Allston. His father, having once made up his mind to allow his son to follow his bent, did everything possible to further his ambition and assist him in his student years. He gave him many letters of introduction to well-known persons in England and France, one of which, to His Excellency C. M. Talleyrand, I shall quote in full.

SIR, — I had the honor to introduce to you, some years since, a young friend of mine, Mr. Wilder, who has since resided in your country. Your civility to him induces me to take the liberty to introduce to you my eldest son, who visits Europe for the purpose of perfecting himself in the art of painting under the auspices of some of your eminent artists. Should he visit France, as he intends, I shall direct him to pay his respects to you, sir, assured that he will receive your protection and patronage so far as you can with convenience afford them.

In thus doing you will much oblige,
Sir, with high consideration
Your most ob'd't. Serv't,
JED. MORSE.

In another letter of introduction, to whom I cannot say, as the address on the copy is lacking, the father says: —

"His parents had designed him for a different profession, but his inclination for the one he has chosen was so strong, and his talents for it, in the opinion of some good judges, so promising, that we thought it not proper to attempt to control his choice.

"In this country, young in the arts, there are few means of improvement. These are to be found in their perfection only in older countries, and in none, perhaps, greater than in yours. In compliance, therefore, with his earnest wishes and those of his friend and patron, Mr. Allston (with whom he goes to London), we have consented to make the sacrifice of feeling (not a small one), and a pecuniary exertion to the utmost of our ability, for the purpose of placing him under the best advantage of becoming eminent in his profession, in hope that he will consecrate his acquisitions to the glory of God and the best good of his fellow men."

Morse arrived in New York on July 6, 1811, after a several days' journey from Charlestown which he describes as very terrible on account of the heat and dust. People were dying from the heat in New York where the thermometer reached 98° in the shade. He says: —

"My ride to New Haven was beyond everything disagreeable; the sun beating down upon the stage (the sides of which we were obliged to shut up on account of the sun) which was like an oven, and the wind, instead of being in our faces as papa supposed, was at our back and brought into our faces such columns of dust as to hinder us from seeing the other side of the stage.

"I never was so completely covered with dust in my life before. Mama, perhaps, will think that I experienced some inconvenience from such a fatiguing journey, but I never felt better in my life than now."

The optimism of youth when it is doing what it wants to do.

He had taken passage on the good ship Lydia with Mr. and Mrs. Allston and some eleven other passengers, and the sailing of the ship was delayed for several days on account of contrary winds, but at last, on July 13, the voyage was begun.

ON BOARD THE LYDIA,
OFF SANDY HOOK, July 15, 1811.

MY DEAR PARENTS, — After waiting a great length of time I have got under way. We left New York Harbor on Saturday, 13th, about twelve o'clock and went as far as the quarantine ground on Staten Island, where, on account of the wind, we waited over Sunday.

We are now under sail with the pilot on board. We have a fair wind from S.S.W. and shall soon be out of sight of land. We have fourteen very agreeable passengers, an experienced and remarkably pleasant captain, and a strong, large, fast-sailing ship. We expect from twenty-five to thirty days' passage. . . . We have a piano-forte on board and two gentlemen who play elegantly, so we shall have fine times. I am in good spirits, though I feel rather singularly to see my native shores disappearing so fast and for so long a time.

I am not yet seasick, but expect to be a little so in a few days. We shall probably be boarded by a British vessel of war soon; there are a number off the coast, but they treat American vessels very civilly.

He kept a careful diary of the voyage to England and again resumed it when he returned to America in 1815. The voyage out was most propitious and lasted but twenty-two days in all: a very short one for that time. As the diary contains nothing of importance relating to the eastern voyage, being simply a record of good weather, fair winds, and pleasant companions, I shall not quote from it at present.

It was all pleasure to the young man, who had never before been away from home, and he sees no reason why people should dread a sea voyage.

The journal of the return trip tells a different story, as we shall see later on, for the passage lasted fifty-seven days, and head winds, gales, and even hurricanes were encountered all the way across, and he wonders why any one should go to sea who can remain safely on land.

LIVERPOOL, August 7, 1811.

MY DEAR PARENTS, — You see from the date that I have at length arrived in England. I have had a most delightful passage of twenty days from land to land and two in coming up the channel.

As this is a letter merely to inform you of my safe arrival I shall not enter into the particulars of our voyage until I get to London, to which place I shall proceed as soon as possible.

Suffice it to say that I have not been sick a moment of the passage, but, on the contrary, have never enjoyed my health better. I have not as yet got my trunks from the custom-house, but presume I shall meet with no difficulty.

I am now at the Liverpool Arms Inn. It is the same inn that Mr. Silliman put up at; it is, however, very expensive; they charge the enormous sum, I believe, of a guinea or a guinea and a half a day.

If I should be detained a day or two in this place I shall endeavor to find out other lodgings; at present, however, it is unavoidable, as all the other passengers are at the same place with me. You may rest assured I shall do everything in my power to be economical, but to avoid imposition of some kind or other cannot be expected, since every one who has been in England and spoken of the subject to me has been imposed upon in some way or other.

You cannot think how many times I have expressed a wish that you knew exactly how I was situated. My passage has been so perfectly agreeable, I know not of a single circumstance that has interfered to render it otherwise, through the whole passage. There has been but one day in which we have not had fair winds. Mr. and Mrs. Allston are perfectly well. She has been sea-sick, but has been greatly benefited by it. She is growing quite healthy. I have grown about three shades darker in consequence of my voyage. I have a great deal to tell you which I must defer till I arrive in London. . . . Oh! how I wish you knew at this moment that I am safe and well in England.

Good-bye. Do write soon and often as I shall.

Your very affectionate son,
SAML. F. B. MORSE.

Everything was new and interesting to the young artist, and his critical observations on people and places, on manners and customs, are naïve and often very keen. The following are extracts from his diary: —

"As to the manners of the people it cannot be expected that I should form a correct opinion of them since my intercourse with them has been so short, but, from what little I have seen, I am induced to entertain a very favorable opinion of their hospitality. The appearance of the women as I met them in the streets struck me on account of the beauty of their complexions. Their faces may be said to be handsome, but their figures are very indifferent and their gait, in walking, is very bad.

"On Friday, the 9th of August, I went to the Mayor to get leave to go to London. He gave me ten days to get there, and told me, if he found me in Liverpool after that time, he should put me in prison, at which I could not help smiling. His name is Drinkwater, but from the appearance of his face I should judge it might be Drinkbrandy.

"On account of his limiting us to ten days we prepared to set out for London immediately as we should be obliged to travel slowly. . . . Mr. and Mrs. Allston and myself ordered a post-chaise, and at twelve o'clock we set out for Manchester, intending to stay there the first night. . . . The people, great numbers of whom we passed, had cheerful, healthy countenances; they were neat in their dress and appeared perfectly happy. . . .

"Much has been said concerning the miserable state in which the lower class of people live in England but especially in large manufacturing cities. That they are so unhappy as some would think I conceive to be erroneous. We are apt to suppose people are unhappy for the reason that, were we taken from our present situation of independence and placed in their situation of depend-ence, we should be unhappy; not considering that contentment is the foundation of happiness. As far as my own observation extends, and from what I can learn on inquiry, the lower class of people generally are contented. N.B. I have altered my opinion since writing this. . . .

"Thus far on our journey we have had a very pleasant time. There is great difference I find in the treatment of travellers. They are treated according to the style in which they travel. If a man arrives at the door of an inn in a stage-coach, he is suffered to alight without notice, and it is taken for granted that common fare will answer for him. But if he comes in a post-chaise, the whole inn is in an uproar; the whole house come to the door, from the landlord down to boots. One holds his hand to help you to alight, another is very officious in showing you to the parlor, and another gets in the baggage, whilst the landlord and landlady are quite in a bustle to know what the gentleman will please to have. This attention, however, is very pleasant, you are sure to be waited upon well and can have everything you will call for, and that of the nicest kind. It is the custom in this country to hire no servants at inns. They, on the contrary, pay for their places and the only wages they get is from the generosity of travellers.

"This circumstance at first would strike a person unacquainted with the customs of England as a very great imposition. I thought so, but, since I have considered the subject better, I believe that there could not be a wiser plan formed. It makes servants civil and obliging and always ready to do anything; for, knowing that they depend altogether on the bounty of travellers, they would fear to do anything which would in the least offend them; and, as there is a customary price for each grade of servants, a person who is travelling can as well calculate the expense of his journey as though they were nothing of the kind."

"*London, August 15, 1811.* You see from the date that I have at length arrived at the place of my destination. I have been in the city about three hours, so you see what is my first object. . . . Mr. and Mrs. Allston with myself took a post-chaise which, indeed, is much more expensive than a stage-coach, but, on account of Mrs. Allston's health, which you know was not very good when in Boston (although she is much benefited by her voyage), we were obliged to travel slowly, and in this manner it has cost us perhaps double the sum which it would have done had we come in a stage-coach. But necessity obliged me to act as I have done. I found myself in a land of strangers, liable to be cheated out of my teeth almost, and, if I had gone to London without Mr. Allston, by waiting at a boarding-house, totally unacquainted with any living creature, I should probably have expended the difference by the time he had arrived. . . . I trust you will not think it extravagant in me for doing as I have done, for I assure you I shall endeavor to be as economical as possible.

"I also mentioned in my letter that I could scarcely expect to steer free from imposition since none of my predecessors have been able to do it. Since writing that letter I have found (in spite of all my care to the contrary) my observation true. In going from the Liverpool Arms to Mr. Woolsey's, which is over a mile, I was under the necessity of getting into a hackney-coach. Upon asking what was to pay he told me a shilling. I offered him half a guinea to change, which I knew to be good, having taken it at the bank in New York.

"He tossed it into the air and caught it in his mouth very dexterously, and, handing it to me back again, told me it was a bad one. I looked at it and told him I was sure it was good, but, appealing to a gentleman who was passing, I found it was bad. Of course I was obliged to give him other money. When I got to my lodgings I related the circumstance to some of my friends and they told me he had cheated me in this way: that it was common for them to carry bad money about them in their mouths, and, when this fellow had caught the good half-guinea in his mouth, he changed it for a bad one. This is one of the thousand tricks they play every day. I have likewise received eleven bad shillings on the road between Liverpool and this place, and it is hardly to be wondered at, for the shilling pieces here are just like old buttons without eyes, without the sign of an impression on them, and one who is not accustomed to this sort of money will never know the difference.

"I find, as mama used to tell me, that I must watch my very teeth or they will cheat me out of them."

"*Friday, 16th, 1811.* This morning I called on Mr. Bromfield and delivered my letters. He received me very cordially, enquired after you particularly, and invited me to dine with him at 5 o'clock, which invitation I accepted. . . . I find I have arrived in England at a very critical state of affairs. If such a state continues much longer, England must fall. American measures affect this country more than you can have any idea of. The embargo, if it had continued six weeks longer, it is said would have forced this country into any measures."

"*Saturday, 17th.* I have been unwell to-day in some degree, so that I have not been able to go out all day. It was a return of the colic. I sent my letter of introduction to Dr. Lettsom with a request that he would call on me, which he did and prescribed a medicine which cured me in an hour or two, and this evening I feel well enough to resume my letter.

"Dr. Lettsom is a very singular man. He looks considerably like the print you have of him. He is a moderate Quaker, but not precise and stiff like the Quakers of Philadelphia. He is a very pleasant and sociable man and withal very blunt in his address. He is a man of excellent information and is considered among the greatest literary characters here. There is one peculiarity, however, which he has in conversation, that of using the verb in the third person singular with the pronoun in the first person singular and plural, as instead of 'I show' or 'we show,' he says 'I shows,' 'we shows,' etc., upon which peculiarity the famous Mr. Sheridan made the following lines in ridicule of him: —

"If patients call, both one and all
I bleeds 'em and I sweats 'em,
And if they die, why what care I —
I. Lettsom.

"This is a liberty I suppose great men take with each other. . . .

"Perhaps you may have been struck at the lateness of the hour set by Mr. Bromfield for dinner [5 o'clock!], but that is considered quite early in London. I will tell you the fashionable hours. A person to be genteel must rise at twelve o'clock, breakfast at two, dine at six, and sup at the same time, and go to bed about three o'clock the next morning. This may appear extravagant, but it is actually practised by the greatest of the fashionables of London. . . .

"I think you will not complain of the shortness of this letter. I only wish you now had it to relieve your minds from anxiety, for, while I am writing, I can imagine mama wishing that she could hear of my arrival, and thinking of thousands of accidents that may have befallen me, and *I wish that in an instant I could communicate the information;* but three thousand miles are not passed over in an instant and we must wait four long weeks before we can hear from each other."

(The italics are mine, for on the outside of this letter written by Morse in pencil are the words: —

"A longing for the telegraph even in this letter.")

"There has a ghost made its appearance a few streets only from me which has alarmed the whole city. It appears every night in the form of shriekings and groanings. There are crowds at the house every night, and, although they all hear the noises, none can discover from whence they come. The family have quitted the house. I suppose 't is only a hoax by some rogue which will be brought out in time."

CHAPTER III

AUGUST 24, 1811 — DECEMBER 1, 1811

Benjamin West. — George III. — Morse begins his studies. — Introduced to West. — Enthusiasms. — Smuggling and lotteries. — English appreciation of art. — Copley. — Friendliness of West. — Elgin marbles. — Cries of London. — Custom in knocking. — Witnesses balloon ascension. — Crowds. — Vauxhall Gardens. — St. Bartholomew's Fair. — Efforts to be economical. — Signs of war. — Mails delayed. — Admitted to Royal Academy. — Disturbances, riots, and murders.

AT this time Benjamin West the American was President of the Royal Academy and at the zenith of his power and fame. Young Morse, admitted at once into the great man's intimacy through his connection with Washington Allston and by letters of introduction, was dazzled and filled with enthusiasm for the works of the master. He considered him one of the greatest of painters, if not the greatest, of all times. The verdict of posterity does not grant him quite so exalted a niche in the temple of Fame, but his paintings have many solid merits and his friendship and favor were a source of great inspiration to the young artist.

Mr. Prime in his biography of Morse relates this interesting anecdote: —

"During the war of American Independence, West, remaining true to his native country, enjoyed the continued confidence of the King, and was actually engaged upon his portrait when the Declaration of Independence was handed to him. Mr. Morse received the facts from the lips of West himself, and communicated them to me in these words: —

"'I called upon Mr. West at his house in Newman

Street one morning, and in conformity with the order given to his servant, Robert, always to admit Mr. Leslie and myself, even if he was engaged in his private studies, I was shown into his studio.

"'As I entered, a half-length portrait of George III stood before me upon an easel, and Mr. West was sitting with back toward me copying from it upon canvas. My name having been mentioned to him, he did not turn, but, pointing with the pencil he had in his hand to the portrait from which he was copying, he said:—

" ' "Do you see that picture, Mr. Morse?"

" ' "Yes sir!" I said; "I perceive it is the portrait of the King."

" ' "Well," said Mr. West, "the King was sitting to me for that portrait when the box containing the American Declaration of Independence was handed to him."

" ' "Indeed," I answered; " what appeared to be the emotions of the King? what did he say?"

" ' "Well, sir," said Mr. West, "he made a reply characteristic of the goodness of his heart," or words to that effect. "'Well, if they can be happier under the government they have chosen than under mine, I shall be happy.' " ' "

On August 24, 1811, Morse writes to his parents:—

"I have begun my studies, the first part of which is drawing. I am drawing from the head of Demosthenes at present, to get accustomed to handling black and white chalk. I shall then commence a drawing for the purpose of trying to enter the Royal Academy. It is a much harder task to enter now than when Mr. Allston was here, as they now require a pretty accurate knowledge of anatomy before they suffer them to enter, and

I shall find the advantage of my anatomical lectures. I feel rather encouraged from this circumstance, since the harder it is to gain admittance, the greater honor it will be should I enter. I have likewise begun a large landscape which, at a bold push, I intend for the Exhibition, though I run the risk of being refused. . . .

"I was introduced to Mr. West by Mr. Allston and likewise gave him your letter. He was very glad to see me, and said he would render me every assistance in his power.

"At the British Institution I saw his famous piece of Christ healing the sick. He said to me: 'This is the piece I intended for America, but the British would have it themselves; but I shall give America the better one.' He has begun a copy, which I likewise saw, and there are several alterations for the better, if it is possible to be better. A sight of that piece is worth a voyage to England of itself. When it goes to America, if you don't go to see it, I shall think you have not the least taste for paintings.

"The encomiums which Mr. West has received on account of that piece have given him new life, and some say he is at least ten years younger. He is now likewise about another piece which will probably be superior to the other. He favored me with a sight of the sketch, which he said he granted to me because I was an American. He had not shown it to anybody else. Mr. Allston was with me and told me afterwards that, however superior his last piece was, this would far exceed it. The subject is Christ before Pilate. It will contain about fifty or sixty figures the size of life.

"Mr. West is in his seventy-sixth year (I think), but,

to see him, you would suppose him only about five-and-forty. He is very active; a flight of steps at the British Gallery he ran up as nimbly as I could. . . . I walked through his gallery of paintings of his own productions; there were upward of two hundred, consisting principally of the original sketches of his large pieces. He has painted in all upwards of six hundred pictures, which is more than any artist ever did with the exception of Rubens the celebrated Dutch painter. . . .

"I was surprised on entering the gallery of paintings in the British Institution, at seeing eight or ten *ladies* as well as gentlemen, with their easels and palettes and oil colors, employed in copying some of the pictures. You can see from this circumstance in what estimation the art is held here, since ladies of distinction, without hesitation or reserve, are willing to draw in public. . . .

"By the way, I digress a little to inform you how I got my segars on shore. When we first went ashore I filled my pockets and hat as full as I could and left the rest in the top of my trunk intending to come and get them immediately. I came back and took another pocket load and left about eight or nine dozen on the top of my clothes. I went up into the city again and forgot the remainder until it was too late either to take them out or hide them under the clothes. So I waited trembling (for contraband goods subject the whole trunk to seizure), but the custom-house officer, being very good-natured and clever, saw them and took them up. I told him they were only for my own smoking and there were so few that they were not worth seizing. 'Oh,' says he, 'I shan't touch them; I won't know they are here,' and then shut down the trunk again. As he smoked, I gave him a couple of dozen for his kindness."

What a curious commentary on human nature it is that even the most pious, up to our own time, can see no harm in smuggling and bribery. And, as another instance of how the standards of right and wrong change with the changing years, further on in this same letter to his strict and pious parents young Morse says: —

"I have just received letters and papers from you by the Galen which has arrived. I was glad to see American papers again. I see by them that the lottery is done drawing. How has my ticket turned out? If the weight will not be too great for one shipload, I wish you would send the money by the next vessel."

The lottery was for the benefit of Harvard College.

"*September 3, 1811.* I have finished a drawing which I intended to offer at the Academy for admission. Mr. Allston told me it would undoubtedly admit me, as it was better than two thirds of those generally offered, but advised me to draw another and remedy some defects in handling the chalks (to which I am not at all accustomed), and he says I shall enter with some éclat. I showed it to Mr. West and he told me it was an extraordinary production, that I had talent, and only wanted knowledge of the art to make a great painter."

In a letter to his friends, Mr. and Mrs. Jarvis, dated September 17, 1811, he says: —

"I was astonished to find such a difference in the encouragement of art between this country and America. In America it seemed to lie neglected, and only thought to be an employment suited to a lower class of people; but here it is the constant subject of conversation, and the exhibitions of the several painters are fashionable resorts. No person is esteemed accomplished or well educated unless he possesses almost an enthusiastic love for paintings. To possess a gallery of pictures is the pride of every nobleman, and they seem to vie with each other in possessing the most choice and most numerous collection. . . . I visited Mr. Copley a few days since. He is very old and infirm. I think his age is upward of seventy, nearly the age of Mr. West. His powers of mind have almost entirely left him; his late paintings are miserable; it is really a lamentable thing that a man should outlive his faculties. He has been a first-rate painter, as you well know. I saw at his room some exquisite pieces which he painted twenty or thirty years ago, but his paintings of the last four or five years are very bad. He was very pleasant, however, and agreeable in his manners.

"Mr. West I visit now and then. He is very liberal to me and gives me every encouragement. He is a very friendly man; he talked with me like a father and wished me to call and see him often and be intimate with him. Age, instead of impairing his faculties, seems rather to have strengthened them, as his last great piece testifies. He is soon coming out with another which Mr. Allston thinks will far surpass even this last. The subject is Christ before Pilate.

"I went last week to Burlington House in Piccadilly, about forty-five minutes' walk, the residence of Lord Elgin, to see some of the ruins of Athens. Lord Elgin has been at an immense expense in transporting the great collection of splendid ruins, among them some of the original statues of Phidias, the celebrated ancient sculptor. They are very much mutilated, however, and impaired by time; still there was enough remaining to show the inferiority of all subsequent sculpture. Even those celebrated works, the Apollo Belvedere, Venus di Medicis, and the rest of those noble statues, must yield to them. . . .

"The cries of London, of which you have doubtless heard, are very annoying to me, as indeed they are to all strangers. The noise of them is constantly in one's ears from morning till midnight, and, with the exception of one or two, they all appear to be the cries of distress. I don't know how many times I have run to the window expecting to see some poor creature in the agonies of death, but found, to my surprise, that it was only an old woman crying 'Fardin' apples,' or something of the kind. Hogarth's picture of the enraged musician will give you an excellent idea of the noise I hear every day under my windows. . . .

"There is a singular custom with respect to knocking at the doors of houses here which is strictly adhered to. A servant belonging to the house rings the bell only; a strange servant knocks once; a market man or woman knocks once and rings; the penny post knocks twice; and a gentleman or lady half a dozen quick knocks, or any number over two. A nobleman generally knocks eight or ten times very loud.

"The accounts lately received from America look rather gloomy. They are thought here to wear a more threatening aspect than they have heretofore done. From my own observation and opportunity of hearing the opinion of the people generally, they are extremely desirous of an amicable adjustment of differences, and seem as much opposed to the idea of war as the better part of the American people. . . .

"In this letter you will perceive all the variety of feeling which I have had for a fortnight past; sometimes in very low, sometimes in very high spirits, and sometimes a balance of each; which latter, though very desirable, I seldom have, but generally am at one extreme or the other. I wrote this in the evenings of the last two weeks, and this will account, and I hope apologize, for its great want of connection."

In a long letter to a friend, dated September 17, 1811, he thus describes some of the sights of London: —

"A few days since I walked about four miles out of town to a village of the name of Hackney to witness the ascension of a Mr. Sadler and another gentleman in a balloon. It was a very grand sight, and the next day the aeronauts returned to Hackney, having gone nearly fifty miles in about an hour and a half. The number of people who attended on this occasion might be fairly estimated at 300,000, such a concourse as I never before witnessed.

"When the balloon was out of sight the crowd began to return home, and such a confusion it is almost impossible for me to describe. A gang of pickpockets had contrived to block up the way, which was across a bridge, with carriages and carts, etc., and as soon as the people began to move it created such an obstruction that, in a few moments, this great crowd, in the midst of which I had unfortunately got, was stopped. This gave the pickpockets an opportunity and the people were plundered to a great amount.

"I was detained in this manner, almost suffocated,

in a great shower of rain, for about an hour, and, what added to the misery of the scene, there were a great many women and children crying and screaming in all directions, and no one able to assist them, not even having a finger at liberty, they were wedged in in such a manner. I had often heard of the danger of a London crowd, but never before experienced it, and I think once is amply sufficient and shall rest satisfied with it.

"A few evenings since I visited the celebrated Vauxhall Gardens, of which you have doubtless often heard. I must say they far exceeded my expectations; I never before had an idea of such splendor. The moment I went in I was almost struck blind with the blaze of light proceeding from thousands of lamps and those of every color.

"In the midst of the gardens stands the orchestra box in the form of a large temple and most beautifully illuminated. In this the principal band of music is placed. At a little distance is another smaller temple in which is placed the Turkish band. On one side of the gardens you enter two splendid saloons illuminated in the same brilliant manner. In one of them the Pandean band is placed, and in the other the Scotch band. All around the gardens is a walk with a covered top, but opening on the sides under curtains in festoons, and these form the most splendid illuminated part of the whole gardens. The amusements of the evening are music, waterworks, fireworks, and dancing.

"The principal band plays till about ten o'clock, when a little bell is rung, and the whole concourse of people (the greater part of which are females) run to a dark part of the gardens where there is an admirable deception of waterworks. A bridge is seen over which stages and wagons, men and horses, are seen passing; birds flying across and the water in great cataracts falling down from the mountains and passing over smaller falls under the bridges; men are seen rowing a boat across, and, indeed, everything which could be devised in such an exhibition was performed.

"This continues for about fifteen minutes, when they all return into the illuminated part of the gardens and are amused by music from the same orchestra till eleven o'clock. They then are called away again to the dark part of the gardens, where is an exhibition of the most splendid fireworks; sky-rockets, serpents, wheels, and fountains of fire in the greatest abundance, occupying twenty minutes more of the time.

"After this exhibition is closed, they again return into the illuminated parts of the gardens, where the music strikes up from the chief orchestra, and hundreds of groups are immediately formed for dancing. Respectable ladies, however, seldom join in this dance, although gentlemen of the first distinction sometimes for amusement lend a hand, or rather a foot, to the general cheerfulness.

"All now is gayety throughout the gardens; every one is in motion, and care, that bane of human happiness, for a time seems to have lost her dominion over the human heart. Had the Eastern sage, who was in search of the land of happiness, at this moment been introduced into Vauxhall, I think his most exalted conceptions of happiness would have been surpassed, and he would rest contented in having at last found the object of his wishes.

"In a few minutes the chief orchestra ceases and is relieved in turn by the other bands, the company following the music. The Scotch band principally plays Scotch reels and dances. The music and this course of dancing continue till about four o'clock in the morning, when the lights are extinguished and the company disperses. On this evening, which was by no means considered as a full night, the company consisted of perhaps three thousand persons.

"I had the pleasure a few days since of witnessing one of the oddest exhibitions, perhaps, in the world. It was no other than *St. Bartholomew's Fair*. It is held here in London once a year and continues three days. There is a ceremony in opening it by the Lord Mayor, which I did not see. At this fair the lower orders of society are let loose and allowed to amuse themselves in any lawful way they please. The fair is held in Smithfield Market, about the centre of the city. The principal amusement appeared to be swinging. There were large boxes capable of holding five or six suspended in large frames in such manner as to vibrate nearly through a semicircle. There were, to speak within bounds, three hundred of these. They were placed all round the square, and it almost made me giddy only to see them all in motion. They were so much pressed for room that one of these swings would clear another but about two inches, and it seemed almost miraculous to me that they did not meet with more accidents than they did.

"Another amusement were large wheels, about thirty or forty feet in diameter, on the circumference of which were four and sometimes six boxes capable of holding four persons. These are set in slow motion, and they gradually rise to the top of the wheel and as gradually descend and so on in succession. There were various other machines on the same principle which I have not time to describe.

"In the centre of the square was an assemblage of everything in the world; theatres, wild beasts, *lusus naturæ*, mountebanks, buffoons, dancers on the slack wire, fighting and swearing, pocket-picking and stealing, music and dancing, and hubbub and confusion in every confused shape.

"The theatres are worth describing; they are temporary buildings put up and ornamented very richly on the exteriors to attract attention, while the interiors, like many persons' heads, are but very poorly furnished. Strolling companies of players occupy these, and between the plays the actors and actresses exhibit themselves on a stage before the theatre in all their spangled robes and false jewels, and strut and flourish about till the theatre is filled.

"Then they go in and turn, perhaps, a very serious tragedy into one of the most ridiculous farces. They occupy about fifteen minutes in reciting a play and then a fresh audience is collected, and so they proceed through the three days and nights, so that the poor actors and actresses are killed about fifty times in the course of a day.

"A person who goes into one of these theatres must not expect to hear a syllable of the tragedy. If he can look upon the stage it is as much as he can expect, for there is such a confused noise without of drums and fifes, clarionets, bassoons, hautboys, triangles, fiddles, bass-viols, and, in short, every possible instrument that can make a noise, that if a person gets safe from the fair without the total loss of his hearing for three weeks he may consider himself fortunate. Contiguous to the theatres are the exhibition rooms of the jugglers and buffoons, who also between their exhibitions display their tricks on stages before the populace, and show as many antics as so many monkeys. But were I to attempt a description of everything I saw at Bartholomew Fair my letter, instead of being a few sheets, would swell to as many quires; so I must close it.

"I shall probably soon witness an exhibition of a more interesting nature; I mean a coronation. The King is now so very low that he cannot survive more than a week or two longer, and immediately on his death the ceremony of the coronation takes place. If I should see it I shall certainly describe it to you."

The King, George III, did not, however, die until 1820.

In a letter of September 20 to his parents he says: "I endeavor to be as economical as possible and am getting into the habit very fast. It must be learned by degrees. I shall not say, as Salmagundi says, — 'I shall spare no expense in discovering the most economical way of spending money,' but shall endeavor to practise it immediately."

"*September 24, 1811.* You will see by the papers which accompany this what a report respecting the capture of the U.S. frigate President by Melampus frigate prevails here. It is sufficient to say it is not in the least credited.

"In case of war I shall be ordered out of the country. If so, instead of returning home, had I not better go to

Paris, as it is cheaper living there even than in London, and there are great advantages there? I only ask the question in case of war. . . . I am going on swimmingly. Next week on Monday the Royal Academy opens and I shall present my drawing."

"*October 21, 1811.* I wrote you by the Galen about three weeks ago and have this moment heard she was still in the Downs. I was really provoked. There is great deception about vessels; they advertise for a certain day and perhaps do not sail under a month after. The Galen has been going and going till I am sick of hearing she has n't gone."

"*November 6, 1811.* After leaving this letter so long, as you see by the different dates, I again resume it. Perhaps you will be surprised when I tell you that but yesterday I heard that the Galen is still wind-bound. It makes my letters which are on board of her about five or six weeks old, besides the prospect of a long voyage. However it is not her fault. There are three or four hundred vessels in the same predicament. The wind has been such that it has been impossible for any of them to get under weigh; but I must confess I feel considerably anxious on your account. . . .

"I mentioned in one of my other letters that I had drawn a figure (the Gladiator) to admit me into the Academy. After I had finished it I was displeased with it, and concluded not to offer it, but to attempt another. I have accordingly drawn another from the Laocoön statue, the most difficult of all the statues; have shown it to the keeper of the Academy and *am admitted for a year* without the least difficulty. Mr. Allston was pleased to compliment me upon it by saying that it was better than two thirds of the drawings of those who had been drawing at the Academy for two years."

"*November 25, 1811.* I mentioned in my last letter that I had entered the Royal Academy, which information I hope will give you pleasure. I now employ my days in painting at home and in the evenings in drawing at the Academy as is customary. I have finished a landscape and almost finished a copy of a portrait which Mr. West lent me. Mr. Allston has seen it and complimented me by saying it was just a hundred times better than he had any idea I could do, and that I should astonish Mr. West very much. I have also begun a landscape, a morning scene at sunrise, which Mr. Allston is very much pleased with. All these things encourage me, and, as every day passes away, I feel increased enthusiasm. . . .

"Distresses are increasing in this country, and disturbances, riots, etc., have commenced as you will see by the papers which accompany this. They are considered very alarming."

"*December 1, 1811.* I am pursuing my studies with increased enthusiasm, and hope, before the three years are out, to relieve you from further expense on my account. Mr. Allston encourages me to think thus from the rapid improvement he says I have made. You may rest assured I shall use all my endeavors to do it as soon as may be. . . .

"This country appears to me to be in a very bad state. I judge from the increasing disturbances at Nottingham, and more especially from the startling murders lately committed in this city.

"A few mornings since was published an account of

the murder of a family consisting of four persons, and this moment there is another account of the murder of one consisting of three persons, making the twelfth murder committed in that part of the city within three months, and not one of the murderers as yet has been discovered, although a reward of more than seven hundred pounds has been offered for the discovery.

"The inhabitants are very much alarmed, and hereafter I shall sleep with pistols at the head of my bed, although there is little to apprehend in this part of the city. Still, as I find many of my acquaintance adopting that plan, I choose rather to be on the safe side and join with them."

CHAPTER IV

JANUARY 18, 1812 — AUGUST 6, 1812

Political opinions. — Charles R. Leslie's reminiscences of Morse, Allston, King, and Coleridge. — C. B. King's letter. — Sidney E. Morse's letter. — Benjamin West's kindness. — Sir William Beechy. — Murders, robberies, etc. — Morse and Leslie paint each other's portraits. — The elder Morse's financial difficulties. — He deprecates the war talk. — The son differs with his father. — The Prince Regent. — Orders in Council. — Estimate of West. — Alarming state of affairs in England. — Assassination of Perceval, Prime Minister. — Execution of assassin. — Morse's love for his art. — Stephen Van Rensselaer. — Leslie the friend and Allston the master. — Afternoon tea. — The elder Morse well known in Europe. — Lord Castlereagh. — The Queen's drawing-room. — Kemble and Mrs. Siddons. — Zachary Macaulay. — Warning letter from his parents. — War declared. — Morse approves. — Gratitude to his parents, and to Allston.

THE years from 1811 to 1815 which were passed by Morse in the study of his art in London are full of historical interest, for England and America were at war from 1812 to 1814, and the campaign of the allied European Powers against Napoleon Bonaparte culminated in Waterloo and the Treaty of Paris in 1815.

The young man took a deep interest in these affairs and expressed his opinions freely and forcibly in his letters to his parents. His father was a strong Federalist and bitterly deprecated the declaration of war by the United States. The son, on the contrary, from his point of vantage in the enemy's country saw things from a different point of view and stoutly upheld the wisdom, nay, the necessity, of the war. His parents and friends urged him to keep out of politics and to be discreet, and he seems, at any rate, to have followed their advice in the latter respect, for he was not in any way molested by the authorities.

At the same time he was making steady progress in his studies and making friends, both among the Americans who were his fellow students or artists of established reputation, and among distinguished Englishmen who were friends of his father.

Among the former was Charles R. Leslie, his roommate and devoted friend, who afterwards became one of the best of the American painters of those days. In his autobiography Leslie says: —

"My new acquaintances Allston, King, and Morse were very kind, but still they were *new* acquaintances. I thought of the happy circle round my mother's fireside, and there were moments in which, but for my obligations to Mr. Bradford and my other kind patrons, I could have been content to forfeit all the advantages I expected from my visit to England and return immediately to America. The two years I was to remain in London seemed, in prospect, an age.

"Mr. Morse, who was but a year or two older than myself, and who had been in London but six months when I arrived, felt very much as I did and we agreed to take apartments together. For some time we painted in one room, he at one window and I at the other. We drew at the Royal Academy in the evening and worked at home in the day. Our mentors were Allston and King, nor could we have been better provided; Allston, a most amiable and polished gentleman, and a painter of the purest taste; and King, warm-hearted, sincere, sensible, prudent, and the strictest of economists.

"When Allston was suffering extreme depression of spirits after the loss of his wife, he was haunted during sleepless nights by horrid thoughts, and he told me that diabolical imprecations forced themselves into his mind. The distress of this to a man so sincerely religious as Allston may be imagined. He wished to consult Coleridge, but could not summon resolution. He desired, therefore, that I should do it, and I went to Highgate where Coleridge was at that time living with Mr. Gillman. I found him walking in the garden, his hat in his hand (as it generally was in the open air), for he told me that, having been one of the Bluecoat Boys, among whom it is the fashion to go bareheaded, he had acquired a dislike to any covering of the head.

"I explained the cause of my visit and he said: 'Allston should say to himself, "*Nothing is me but my will.* These thoughts, therefore, that force themselves on my mind are no part of *me* and there can be no guilt in them." If he will make a strong effort to become indifferent to their recurrence, they will either cease or cease to trouble him.'

"He said much more, but this was the substance, and, after it was repeated to Allston, I did not hear him again complain of the same kind of disturbance."

Mr. C. B. King, the other friend mentioned by Leslie, returned to America in 1812, and writes from Philadelphia, January 3, 1813: —

My dear Friends, This will be handed you by Mr. Payne, of Boston, who intends passing some time in England. . . . I have not been here sufficiently long to forget the delightful time when we could meet in the evening with novels, coffee, and *music by Morse*, with the conversation of that dear fellow Allston. The reflection that it will not again take place, comes across my mind accompanied with the same painful sensation as the thought that I must die.

That Morse was not forgotten by the good people at home is evidenced by a letter from his brother, Sidney Edwards, of January 18, 1812, part of which I transcribe: —

Dear Brother, — I am sitting in the parlor in the armchair on the right of the fireplace, and, as I hold my paper in my hand, with my feet sprawled out before the fire, and with my body reclining in an oblique position against the back of the chair, I am penning you a letter such as it is, and for the inverted position of the letters of which I beg to apologize.

As I turn my eyes upward and opposite I behold the family picture painted by an ingenious artist who, I understand, is at present residing in London. If you are acquainted with him, give my love to him and my best wishes for his prosperity and success in the art to which, if report says true, he has devoted himself with much diligence.

Richard sits before me writing to you, and mama says (for I have just asked her the question) that she is engaged in the same business. Papa is upstairs very much engaged in the selfsame employment. Four right hands are at this instant writing to give you, at some future moment, the pleasure of perusing the products of their present labor. Four imaginations are now employed in conceiving of a son or a brother in a distant land. Therefore we may draw the conclusion that you are not universally forgotten, and consequently all do not forget you.

I have written you this long letter because I knew that you would be anxious for the information it contains; because papa told me I must write; because mama said I had better write; because I had nothing else to do, and because I had n't time to write a shorter. I trust for these special reasons you will excuse me for this once, especially when you consider that you asked me to write you long letters; when you consider that it is my natural disposition to express my sentiments fully; that I commonly say most when I have least *to* say; that I promise reformation in future, and that you shall hereafter hear from me on this subject.

As to news, I am sorry to say we are entirely out. We sent you the last we had by the Sally Ann. We hope to get some ready by the time the next ship sails, and then we will furnish you with the best the country affords.

From a letter of January 30, 1812, to his parents I select the following passages: —

"On Tuesday last I dined at Mr. West's, who requested to be particularly remembered to you. He is extremely attentive and polite to me. He called on me a few days ago, which I consider a very marked attention as he keeps so confined that he seldom pays any visits. . . .

"I have changed my lodgings to No. 82 in the same street [Great Titchfield Street], and have rooms with young Leslie of Philadelphia who has just arrived. He is very promising and a very agreeable room-mate. We are in the same stage of advancement in art.

"I have painted five pieces since I have been here, two landscapes and three portraits; one of myself, one a copy from Mr. West's copy from Vandyke, and the other a portrait of Mr. Leslie, who is also taking mine. . . . I called a day or two since on Sir William Beechy, an artist of great eminence, to see his paintings. They are beautiful beyond anything I ever imagined. His principal excellence is in coloring, which, to the many, is the most attractive part of art. Sir William is considered the best colorist now living.

"You may be apt to ask, 'If Sir William is so great and even the best, what is Mr. West's great excellence?' Mr. West is a bad colorist in general, but he excels in the grandeur of his thought. Mr. West is to painting what Milton is to poetry, and Sir William Beechy to Mr. West as Pope to Milton, so that by comparing, or rather illustrating the one art by the other, I can give you a better idea of the art of painting than in any other way. For as some poets excel in the different species of poetry and stand at the head of their different kinds, in the same manner do painters have their particular branch of their art; and as epic poetry excels all other kinds of poetry, because it addresses itself to the sublimer feelings of our nature, so does historical painting stand preeminent in our art, because it calls forth the same feelings. For poets' and painters' minds are the same, and I infer that painting is superior to poetry from this: — that the painter possesses with the poet a vigorous imagination, where the poet stops, while the painter exceeds him in the mechanical and very difficult part of the art, that of handling the pencil.

"I gave you a hint in letter number 12 and a particular account in number 13 of the horrid murders committed in this city. It has been pretty well ascertained from a variety of evidence that all of them have been committed by one man, who was apprehended and put an end to his life in prison. Very horrid attempts at robbery and murder have been very frequent of late in all parts of the city, and even so near as within two doors of me in the same street, but do not be alarmed, you have nothing to fear on my account. Leslie and myself sleep in the same room and sleep armed with a pair of pistols and a sword and alarms at our doors and windows, so we are safe on that score. . . .

"In my next I shall give you some account of politics here and as it respects America. The Federalists are certainly wrong in very many things. . . .

"P.S. I wish you would keep my letter in which I enumerate all my friends, and when I say, 'Give my love to my friends,' imagine I write them all over, and distribute it out to all as you think I ought, always particularizing Miss Russell, my patroness, my brothers, relations, and Mr. Brown and Nancy [his old nurse]. This will save me time, ink, trouble, and paper."

Concerning the portraits which Morse and Leslie were painting of each other, the following letter to Morse's mother, from a friend in Philadelphia and signed "R. W. Snow," will be found interesting: —

My dear Friend, — I have this moment received a letter from Miss Vaughan in London, dated February 20, 1812, and, knowing the passage below would be interesting to you, I transcribe it with pleasure, and add my very sincere wish that all your hopes may be realized.

"Dr. Morse's son is considered a young man of very promising talents by Mr. Allston and Mr. West and by those who have seen his paintings. We have seen him and think his modesty and apparent amiableness promise as much happiness to his friends as his talents may procure distinction for himself. He is peculiarly fortunate, not only in having Mr. Allston for an adviser and friend, but in his companion in painting, Mr. Leslie, a young man from Philadelphia highly recommended by my uncle there, and whose extreme diffidence adds to the most promising talents the patient industry and desire of improvement which are necessary to bring them to perfection. They have been drawing each other's pictures. Mr. Leslie is in the Spanish costume and Mr. Morse in Highland dress. They are in an unfinished state, but striking resemblances."

This Highland lad, I hope, my dear friend, you will see, and in due time be again blessed with the interesting original.

At this time the good father was sore distressed financially. He was generous to a fault and had, by endorsing notes and giving to others, crippled his own means. He says in a letter to his son dated March 21, 1812: —

"The Parkman case remains yet undecided and I know not that it ever will be. There is a strange mystery surrounding the business which I am not able to unravel. The court is now in session in Boston which is expected to decide the case. In a few days we shall be able to determine what we have to expect from this case. If we lose it, your mother and I have made up our minds to sit down contented with the loss. I trust we shall be enabled to pay our honest debts without it and to support ourselves.

"As to you and your brothers, I trust, with your education, you will be able to maintain yourselves, and your parents, too, should they need it in their old age. Probably this necessity laid on you for exertion, industry, and economy in early life will be better for you in the end than to be supported by your parents. In nine cases out of ten those who begin the world with nothing are richer and more useful men in life than those who inherit a large estate. . . .

"We have just heard from your brothers, who are well and in fine spirits. Edwards writes that he thinks of staying in New Haven another year and of pursuing *general science*, and afterwards of purchasing a plantation and becoming a planter in some one of the Southern States!! Perhaps he intends to marry some rich planter's daughter and to get his plantation and negroes in that way. This, I imagine, will be his only way to do it.

"The newspapers which I shall send with this will inform you of the state of our public affairs. We have high hopes that Governor Strong will be our governor next year. I have no belief that our *war hawks* will be able to involve the country in a war with Great Britain, nor do I believe that the President really wishes it. It is thought that all the war talk and preparations are intended to effect the reëlection of Mr. Madison. The *Henry Plot* is a farce intended for the same purpose, but it can never be got up. It will operate against its promoters."

While the father was thus writing, on March 21, of the political conditions in America from his point of view, almost at the same moment the son in England was expressing himself as follows: —

"*March 25, 1812.* With respect to politics I know very little, my time being occupied with much pleasanter subjects. I, however, can answer your question whether party spirit is conducted with such virulence here as in America. It is by no means the case, for, although it is in some few instances very violent, still, for the most part, their debates are conducted with great coolness.

"As to the Prince Regent, you have, perhaps, heard how unpopular he has made himself. He has disappointed the expectations of very many. Among the most unpopular of his measures may be placed the retention of the Orders in Council, which orders, notwithstanding the declarations of Mr. Perceval [the Prime Minister] and others in the Ministry to the contrary, are fast, very fast reducing this country to ruin; and it is the opinion of some of the best politicians in this country that, should the United States either persist in the Non-Intercourse Law or declare war, this country would be reduced to the lowest extremity.[1]

"Bankruptcies are daily increasing and petitions from all parts of the Kingdom, praying for the repeal of the Orders in Council, have been presented to the Prince, but he has declined hearing any of them. Also the Catholic cause remains undecided, and he refuses hearing anything on that subject. But no more of politics. I am sure you must have more than sufficient at home.

"I will turn to a more pleasant subject and give you a slight history of the American artists now in London.

[1] Orders in Council were issued by the sovereign, with the advice of the Privy Council, in periods of emergency, trusting to their future ratification by Parliament. In this case, while promulgated as a retaliatory measure against Bonaparte's Continental System, they bore heavily upon the commerce of the United States.

"At the head stands Mr. West. He stands and has stood so long preëminent that I could relate but little of his history that would be new to you, so that I shall confine myself only to what has fallen under my own observation, and, of course, my remarks will be few.

"As a painter Mr. West can be accused of as few faults as any artist of ancient or modern times. In his studies he has been indefatigable, and the result of those studies is a perfect knowledge of the philosophy of his art. There is not a line or a touch in his pictures which he cannot account for on philosophical principles. They are not the productions of accident, but of study.

"His principal excellence is considered composition, design, and elegant grouping; and his faults were said to be a hard and harsh outline and bad coloring. These faults he has of late in a great degree amended. His outline is softer and his coloring, in some pictures in which he has attempted truth of color, is not surpassed by any artist now living, and some have even said that Titian himself did not surpass it. However that may be, his pictures of a late date are admirable even in this particular, and it evinces that, if in general he neglected that fascinating branch of art in some of his paintings, he still possesses a perfect knowledge of all its artifices. He has just completed a picture, an historical landscape, which, for clearness of coloring combined with grandeur of composition, has never been excelled.

"In his private character he is unimpeachable. He is a man of tender feelings, but of a mind so noble that it soars above the slanders of his enemies, and he expresses pity rather than revenge towards those who, through wantonness or malice, plan to undermine his character. No man, perhaps, ever passed through so much abuse, and none, I am confident, ever bore up against its virulence with more nobleness of spirit, with a steady perseverance in the pursuit of the sublimest of human professions. He has travelled on heedless of the sneers, the ridicule, or the detraction of his enemies, and he has arrived at that point where the lustre of his works will not fail to illuminate the dark regions of barbarism and distaste long after their bright author has ceased to exist.

"Excuse my fervor in the praise of this man. He is not a common man, not such a one as can be met with in every age. He is one of those geniuses who are doomed in their lifetime to endure the malice, the ridicule, and neglect of the world, and at their death to receive the praise and adoration of this same inconsistent world. I think there cannot be a stronger proof that human nature is always the same than that men of genius in all ages have been compelled to undergo the same disappointments and to pass through the same routine of calumny and abuse."

The rest of this letter is missing, which is a great pity, as it would be interesting to read what Morse had to say of Allston, Leslie, and the others.

Was it a presentiment of the calumnies and abuse to which he himself was to be subjected in after life which led him to express himself so heartily in sympathy with his master West? And was it the inspiring remembrance of his master's calm bearing under these afflictions which heartened him to maintain a noble serenity under even greater provocation?

"*April 21, 1812.* I mentioned in my last letter that I should probably exceed my allowance this year by a few pounds, but I now begin to think that I shall not. I am trying every method to be economical and hope it will not be long before I shall relieve you from further expense on my account. . . .

"With respect to politics they appear gloomy on both sides. . . . You may depend on it England has injured us sorely and our Non-Intercourse is a just retaliation for those wrongs. Perhaps you will believe what is said in some of the Federal papers that that measure has no effect on this country. You may be assured the effects are great and severe; I am myself an eye-witness of the effects. The country is in a state of rebellion from literal starvation. Accounts are daily received which grow more and more alarming from the great manufacturing towns. Troops are in motion all over the country, and but last week measures were adopted by Parliament to prevent this metropolis from rising to rebellion, by ordering troops to be stationed round the city to be ready at a moment's warning. This I call an alarming period. Everybody thinks so and Mr. Perceval himself is frightened, and a committee is appointed to take into consideration the Orders in Council. Now, when you consider that I came to this country prejudiced against our government and its measures, and that I can have no bad motive in telling you these facts, you will not think hard of me when I say that I hope that our Non-Intercourse Law will be enforced with all its rigor, as I firmly believe it is the only way to bring this country to terms, and that, if persisted in, it will certainly bring them to terms. I know it must make some misery at home, but it will be followed by a corresponding happiness after it. Some of you at home, I suppose, will call me a Democrat, but facts are stubborn things, and I can't deny the truth of what I see every day before my eyes. A man to judge properly of his country must, like judging of a picture, view it at a distance."

"*May 12, 1812.* I write in great haste to inform you of a dreadful event which happened here last evening, and rumors of which will probably reach you before this. Not to keep you in suspense it is no less than the *assassination* of *Mr. Perceval,* the Prime Minister of Great Britain. As he was entering the House of Commons last evening a little past five o'clock, he was shot directly through the heart by a man from behind the door. He staggered forward and fell, and expired in about ten minutes. . . .

"I have just returned from the House of Commons; there was an immense crowd assembled and very riotous. In the hall was written in large letters, 'Peace or the Head of the Regent.' This country is in a very alarming state and there is no doubt but great quantities of blood will be spilled before it is restored to order. Even while I am writing a party of Life Guards is patrolling the streets. London must soon be the scene of dreadful events.

"Last night I had an opportunity of studying the public mind. It was at the theatre; the play was 'Venice Preserved; or, the Plot Discovered.' If you will take the trouble just to read the first act you will see what relation it has to the present state of affairs. When Pierre says to Jaffier, 'Cans't thou kill a Senator?' there were three cheers, and so through the whole, whenever anything was said concerning conspiracy and in favor of it, the audience applauded, and when anything was said against it they hissed. When Pierre asked the conspirators if Brutus was not a good man, the audience was in a great uproar, applauding so as to prevent for some minutes the progress of the performance. This I think shows the public mind to be in great agitation. The play of 'Venice Preserved' is not a moral play, and I should not ask you to read any part of it if I could better explain to you the feelings of the public."

A few days later, on May 17, he says in a letter to his brothers: —

"The assassin Bellingham was immediately taken into custody. He was tried on Friday and condemned to be executed to-morrow morning (Monday, 18th). I shall go to the place to see the concourse of people, for to see him executed I know I could not bear."

In a postscript written the day after he says: —

"I went this morning to the execution. A very violent rain prevented so great a crowd as was expected. A few minutes before eight o'clock Bellingham ascended the scaffold. He was very genteelly dressed; he bowed to the crowd, who cried out, 'God bless you,' repeatedly. I saw him draw the cap over his face and shake hands with the clergyman. I stayed no longer, but immediately turned my back and was returning home. I had taken but a few steps when the clock struck eight, and, on turning back, I saw the crowd beginning to disperse. I have felt the effects of this sight all day, and shall probably not get over it for weeks. It was a dreadful sight. There were no accidents."

In spite of all these momentous occurrences, the

young artist was faithfully pursuing his studies, for in this same letter to his brothers he says: —

"But enough of this; you will probably hear the whole account before this reaches you. I am wholly absorbed in the studies of my profession; it is a slow and arduous undertaking. I never knew till now the difficulties of art, and no one can duly appreciate it unless he has tried it. Difficulties, however, only increase my ardor and make me more determined than ever to conquer them.

"Mr. West is very kind to me; I visit him occasionally of a morning to hear him converse on art. He appears quite attached to me, as he is, indeed, to all young American artists. It seems to give him the greatest pleasure to think that one day the arts will flourish in America. He says that Philadelphia will be the Athens of the world. That city certainly gives the greatest encouragement of any place in the United States. Boston is most backward, so, if ever I should return to America, Philadelphia or New York would probably be my place of abode.

"I have just seen Mr. Stephen Van Rensselaer, who you know was at college with us, and with whom I was intimate. He was very glad to see me and calls on me every day while I am painting. He keeps his carriage and horses and is in the first circles here. I ride out occasionally with him; shall begin his portrait next week."

Like a breath of fresh air, in all the heat and dust of these troublous times, comes this request from his gentle mother in a letter of May 8, 1812: —

"Miss C. Dexter requests the favor of you to take a sketch of the face of Mr. Southey and send it her.

He is a favorite writer with her and she has a great desire to see the style of his countenance. If you can get it, enclose it in a genteel note to her with a brief account of him, his age and character, etc."

The next letter of May 25, 1812, is from Morse to his parents.

"I have told you in former letters that my lodgings are at 82 Great Titchfield Street and that my room-mate is Leslie, the young man who is so much talked of in Philadelphia. We have lived together since December and have not, as yet, had a falling out. I find his thoughts of art agree perfectly with my own. He is enthusiastic and so am I, and we have not time, scarcely, to think of anything else; everything we do has a reference to art, and all our plans are for our mutual advancement in it. Our amusements are walking, *occasionally* attending the theatres, and the company of Mr. Allston and a few other gentlemen, consisting of three or four painters and poets. We meet by turn at each other's rooms and converse and laugh.

"Mr. Allston is our most intimate friend and companion. I can't feel too grateful to him for his attentions to me; he calls every day and superintends all we are doing. When I am at a stand and perplexed in some parts of the picture, he puts me right and encourages me to proceed by praising those parts which he thinks good, but he is faithful and always tells me when anything is bad.

"It is a mortifying thing sometimes to me, when I have been painting all day very hard and begin to be pleased with what I have done, on showing it to Mr. Allston, with the expectation of praise, and not only of praise but a score of 'excellents,' 'well dones,' and 'admirables'; I say it is mortifying to hear him after a long silence say: 'Very bad, sir; that is not flesh, it is mud, sir; it is painted with brick dust and clay.'

"I have felt sometimes ready to dash my palette knife through it and to feel at the moment quite angry with him; but a little reflection restores me; I see that Mr. Allston is not a flatterer but a friend, and that really to improve I must see my faults. What he says after this always puts me in good humor again. He tells me to put a few flesh tints here, a few gray ones there, and to clear up such and such a part by such and such colors. And not only that, but takes the palette and brushes and shows me how, and in this way he assists me. I think it one of the greatest blessings that I am under his eye. I don't know how many errors I might have fallen into if it had not been for his attentions. . . .

"I am painting portraits alone at present. Our sitters are among our acquaintances. We paint them if they defray the expense of canvas and colors. . . .

"Mama wished me to send some specimens of my painting home that you might see my improvement. The pictures that I now paint would be uninteresting to you; they consist merely of studies and drawings from plaster figures, hands and feet and such things. The portraits are taken by those for whom they are painted. I shall soon begin a portrait of myself and will try and send that to you."

"*June 8, 1812.* Mama asks in one of her letters if we make our own tea. We do. The tea-kettle is brought to us boiling in the morning and evening and we make our own coffee (which, by the way, is very cheap here) and tea. We live quite in the old bachelor style. I don't know but it will be best for me to live in this style through life; my profession seems to require all my time.

"Mr. Hurd will take a diploma to you, with others to different persons near Boston. I suppose it confers some title on you of consequence, as I saw at his house a great number to be sent to all parts of the world to distinguished men. I find papa is known here pretty extensively. Some one, hearing my name and that I am an American, immediately asks if I am related to you. . . .

"The Administration is at length formed, and, to the great sorrow of everybody, the old Ministers are reelected. The Orders in Council are the subject of debate at the House of Commons this evening. It is an important crisis, though there is scarcely any hope of their repeal. If not, I sincerely hope that America will declare war.

"What Lord Castlereagh said at a public meeting a few days ago ought to be known in America. Respecting the Orders in Council, when some one said unless they were repealed war with America must be the consequence, he replied that, '*if the people would but support the Ministry in those measures for a short time, America would be compelled to submit, for she was not able to go to war.*' But I say, and so does every American here who sees how things are going with this country, that, should America but declare war, before hostilities commenced Great Britain would sue for peace on any terms. Great Britain is jealous of us and would trample on us if she could, and I feel ashamed when I see her supported through everything by some of the Federal editors. I wish they could be here a few months and they would be ashamed of themselves. They are injuring their country, for it is *their* violence that induces this Government to persist in their measures by holding out hope that the parties will change, and that then they can compel America to do anything. If America loses in this contest and softens her measures towards this country, she never need expect to hold up her head again."

"*June 15, 1812.* The Queen held a drawing-room a short time since and I went to St. James's Palace to see those who attended. It was a singular sight to see the ladies and gentlemen in their court dresses. The gentlemen were dressed in buckram skirted coats without capes, long waistcoats, cocked hats, bag-wigs, swords, and large buckles on their shoes. The ladies in monstrous hoops, so that in getting into their carriages they were obliged to go edgewise. Their dresses were very rich; some ladies, I suppose, had about them to adorn them £20,000 or £30,000 worth of diamonds.

"I had a sight of the Prince Regent as he passed in his splendid state carriage drawn by six horses. He is very corpulent, his features are good, but he is very red and considerably bloated. I likewise saw the Princess Charlotte of Wales, who is handsome, the Dukes of Kent, Cambridge, Clarence, and Cumberland, Admiral Duckworth, and many others. The Prince held a levee a few days since at which Mr. Van Rensselaer was presented.

"I occasionally attend the theatres. At Covent Garden there is the best acting in the world; Mr. Kemble is the first tragic actor now in England; Cook was a rival and excelled him in some characters. Mrs. Siddons is the first tragic actress, perhaps, that ever lived. She is now advanced in life and is about to retire from the stage; on the 29th of this month she makes her last appearance. I must say I admire her acting very much; she is rather corpulent, but has a remarkably fine face; the Grecian character is finely portrayed in it; she excels to admiration in deep tragedy. In Mrs. Beverly, in the play of the 'Gamesters' a few nights ago, she so arrested the attention of the house that you might hear your watch tick in your fob, and, at the close of the play, when she utters an hysteric laugh for joy that her husband was not a murderer, there were different ladies in the boxes who actually went into hysterics and were obliged to be carried out of the theatre. This I think is proof of good acting. Mrs. Siddons is a woman of irreproachable character and moves in the first circles; the stage will never again see her equal.

"You must n't think because I praise the acting that I am partial to theatres. I think in a certain degree they are harmless, but, too much attended, they dissipate the mind. There is no danger of my loving them too much; I like to go once in awhile after studying hard all day.

"Last night, as I was passing through Tottenham Court Road, I saw a large collection of people of the lower class making a most terrible noise by beating on something of the sounding genus. Upon going nearer and enquiring the cause, I found that a butcher had just been married, and that it is always the custom on such occasions for his brethren by trade to serenade the couple with *marrow-bones* and *cleavers*. Perhaps you have heard of the phrase 'musical as marrow-bones and cleavers'; this is the origin of it. If you wish to experience the sound let each one in the family take a pair of tongs and a shovel, and then, standing all together, let each one try to outdo the other in noise, and this will give you some idea of it. How this custom originated I don't know. I hope it is not symbolical of the *harmony* which is to exist between the parties married."

Among those eminent Englishmen to whom young Morse had letters of introduction was Zachary Macaulay, editor of the "Christian Observer," and father of the historian. The following note from him will be found of a delightful old-time flavor: —

Mr. Macaulay presents his compliments to Mr. Morse and begs to express his regret at not having yet been so fortunate as to meet with him. Mr. Macaulay will be particularly happy if it should suit Mr. Morse to dine with him at his house at Clapham on Saturday next at five o'clock. Mr. M.'s house is five doors beyond the Plough at the entrance of Clapham Common. A coach goes daily to Clapham from the Ship at Charing Cross at a quarter past three, and several leave Grace Church Street in the City every day at four. The distance from London Bridge to Mr. Macaulay's house is about four miles.

23d June, 1812.

In a letter from his mother of June 28, 1812, the anxious parent says: —

"Although we long to see you, yet we rejoice that you are so happily situated at so great a distance from our, at present, wretched, miserably distracted country, whose mad rulers are plunging us into an unnecessary war with a country that I shall always revere as doing more to spread the glorious gospel of Jesus Christ to the benighted heathen, and those that are famishing from lack of knowledge, than any other nation on the globe. Our hearts bleed at every pore to think of again being at war. We have not yet forgotten the wormwood and gall of the last revolution.

"We hope you will steer clear of any of the difficulties of the contest that is about to take place. We wish you to be very prudent and guarded in all your conversation and actions and not to make yourself a party man on either side. Have your opinions, but have them to yourself, and be sure you do not commit them to paper. It may do you great injury either on one side or the other, and you are not in your present situation as a politician but as an artist."

In this same letter his father adds: —

"The die is cast and our country plunged in war. . . . There is great opposition to it in the country. The papers, which you will have opportunity to see, will inform you of the state of parties. Your mother has given you sound advice as respects the course you should pursue. Be the *artist* wholly and let *politics* alone. I rejoice that you are where you are at the present time. You will do what you can without delay to support yourself, as I know not how we shall be able to procure funds to transmit to you, and, if we had them, how we could transmit them should the war continue."

To this the son answers in a letter of August 6, 1812: —

"I am improving, perhaps, the last opportunity I shall have for some time to write you. Mr. Wheeler, an American, who has been here some time studying portrait painting, has kindly offered to deliver this to you.

"Our political affairs, it seems, have come to a crisis, which I sincerely hope will turn to the advantage of America; it certainly will not to this country. War is an evil which no man ought to think lightly of, but, if ever it was just, it now is. The English acknowledge it, and what can be more convincing proof than the confession of an enemy? I was sorry to hear of the riotous proceedings in Boston. If they knew what an injury they were doing their country in the opinion of foreign nations, they certainly would refrain from them. I assert (because I have proof) that the Federalists in the Northern States have done more injury to their country by their violent opposition measures than even a French alliance could. Their proceedings are copied into the English papers, read before Parliament, and circulated through the country, and what do they say of them? Do they say the Federalists are patriots and are firm in asserting the rights of their country? No; they call them *cowards*, a *base set;* say they are traitors to their country and ought to be hanged like traitors. These things I have heard and read, and therefore must believe them.

"I wish I could have a talk with you, papa; I am sure I could convince you that neither Federalists nor Democrats are Americans; that war with this country is just, and that the present Administration of our country has acted with perfect justice in all their proceedings against this country. . . .

"To observe the contempt with which America is spoken of, and the epithets of a '*nation of cheats,*' '*sprung from convicts,' 'pusillanimous,' 'cowardly,'* and such like, — these I think are sufficient to make any true American's blood boil. These are not used by individuals only, but on the floor of the House of Commons. The good effects of our declaration of war begin to be perceived already. The tone of their public prints here is a little softer and more submissive. Not one has called in question the justice of the declaration of war; all say, 'We are in the wrong and we shall do well to get out of it as soon as possible.'

"I could tell you volumes, but I have not time, and it would, perhaps, be impolitic in the present state of affairs. I only wish that among the infatuated party men I may not find my father, and I hope that he will be *neutral* rather than oppose the war measure, for (if he will believe a son who loves him and his country better the longer and farther he is away from them) this war will reëstablish that character for honor and spirit which our country has lost through the proceedings of *Federalists.*

"But I will turn from this subject. My health and spirits are excellent and my love for my profession increases. I am painting a small historical piece; the subject is 'Marius in Prison,' and the soldier sent to kill him who drops his sword as Marius says, '*Durst thou kill Caius Marius?*' The historical fact you must be familiar with. I am taking great pains with it, and may possibly exhibit it in February at the British Gallery.

"I never think of my situation in this country but with gratitude to you for suffering me to pursue the profession of my choice, and for making so many sacrifices to gratify me. I hope I shall always feel grateful to the best of parents and be able soon to show them I am so. In the mean time, if industry and application on my part can make them happy, be assured I shall use my best endeavors to be industrious, and in any other way to give them comfort. One of my greatest blessings here is Mr. Allston. He is like a brother to me, and not only is a most agreeable and entertaining companion, but he has been the means of giving me more knowledge (practical as well as theoretical) in my art than I could have acquired by myself in three years.

"In whatever circumstance I am, Mr. Allston I shall esteem as one of my best and most intimate friends, and in whatever I can assist him or his I shall feel proud in being able to do it.

"Mr. and Mrs. Allston are well. I dined with them yesterday at Captain Visscher's, whom I have mentioned to you before as one of our passengers. He is very attentive to us, visits us constantly, and is making us presents of various kinds every day, such as half a dozen best Madeira, etc. He came out here with his lady to take possession of a fortune of £80,000 and was immensely rich before, having married Miss Van Renselaer of Albany."

CHAPTER V

SEPTEMBER 20, 1812 — JUNE 13, 1813

Models the "Dying Hercules." — Dreams of greatness. — Again expresses gratitude to his parents. — Begins painting of "Dying Hercules." — Letter from Jeremiah Evarts. — Morse upholds righteousness of the war. — Henry Thornton. — Political discussions. — Gilbert Stuart. — William Wilberforce. — James Wynne's reminiscences of Morse, Coleridge, Leslie, Allston, and Dr. Abernethy. — Letters from his mother and brother. — Letters from friends on the state of the fine arts in America. — "The Dying Hercules" exhibited at the Royal Academy. — Expenses of painting. — Receives Adelphi Gold Medal for statuette of Hercules. — Mr. Dunlap's reminiscences. — Critics praise "Dying Hercules."

THE young artist's letters to his parents at this period are filled with patriotic sentiments, and he writes many pages descriptive of the state of affairs in England and of the effects of the war on that country. He strongly upholds the justice of that war and pleads with his parents and brothers to take his view of the matter. They, on the other hand, strongly disapprove of the American Administration's position and of the war, and are inclined to censure and to laugh at the enthusiastic young man's heroics.

As we are more concerned with Morse's career as an artist than with his political sentiments, and as these latter, I fear, had no influence on the course of international events, I shall quote but sparingly from that portion of the correspondence, just enough to show that, whatever cause he espoused, then, and at all times during his long life, he threw himself into it heart and soul, and thoroughly believed in its righteousness. He was absolutely sincere, although he may sometimes have been mistaken.

In a letter dated September 20, 1812, he says: —

"I have just finished a model in clay of a figure (the 'Dying Hercules'), my first attempt at sculpture. Mr. Allston is extremely pleased with it; he says it is better than all the things I have done since I have been in England put together, and says I must send a cast of it home to you, and that it will convince you that I shall make a painter. He says also that he will write to his friends in Boston to call on you and see it when I send it.

"Mr. West also was extremely delighted with it. He said it was not merely an academical figure, but displayed mind and thought. He could not have made me a higher compliment.

"Mr. West would write you, but he has been disabled from painting or writing for a long time with the gout in his right hand. This is a great trial to him.

"I am anxious to send you something to show you that I have not been idle since I have been here. My passion for my art is so firmly rooted that I am confident no human power could destroy it. [And yet, as we shall see later on, human injustice so discouraged him that he dropped the brush forever.]

"The more I study it, the greater I think is its claim to the appellation of '*divine*,' and I never shall be able sufficiently to show my gratitude to my parents for their indulgence in so greatly enabling me to pursue that profession, without which I am sure I would be miserable. If ever it is my destiny to become great and worthy of a biographical memoir, my biographer will never be able to charge upon my parents that bigoted attachment to any individual profession, the exercise of which spirit by parents toward their children has been the ruin of some of the greatest genuises; and the biography of men of genius has too often contained that reflection on their parents. If ever the contrary spirit was evident, it has certainly been shown by my parents towards me. Indeed, they have been almost too indulgent; they have watched every change of my capricious inclinations, and seem to have made it an object to study them with the greatest fondness. But I think they will say that, when my desire for change did cease, it always settled on painting.

"I hope that one day my success in my profession will reward you, in some measure, for the trouble and inconvenience I have so long put you to.

"I am now going to begin a picture of the death of Hercules from this figure, as large as life. The figure I shall send to you as soon as it is practicable, and also one of the same to Philadelphia, if possible in time for the next exhibition in May.

"I have enjoyed excellent health and spirits and am perfectly contented. The war between the two countries has not been productive of any measures against resident American citizens. I hope it will produce a good effect towards both countries."

He adds in a postscript that he has removed from 82 Great Titchfield Street to No. 8 Buckingham Place, Fitzroy Square.

The following extract from a letter to Morse written by his friend, Mr. Jeremiah Evarts, father of William M. Evarts, dated Charlestown, October 7, 1812, is interesting: —

"I am happy that you are so industriously and prosperously engaged in the prosecution of your profession. I hope you will let politics entirely alone for many reasons, not the least of which is a regard to the internal tranquillity of your own mind. I never yet knew a man made happy by studying politics; nor useful, unless he has great duties to perform as a citizen. You will receive this advice, I know, with your accustomed good nature."

The next letter, dated November 1, 1812, is a very long one, over eighteen large pages, and is an impassioned appeal to his father to look at the war from the son's point of view. I shall quote only a few sentences.

"Your last letter was of October 2, via Halifax, accompanying your sermon on Fast Day. The letter gave me great pleasure, but I must confess that the sentiments in the sermon appeared very *strange* to me, knowing what I, as well as every American here does, respecting the causes of the present war. . . . 'T is the character of Englishmen to be haughty, proud, and overbearing. If this conduct meets with no resistance, their treatment becomes more imperious, and the more submissive and conciliating is the object of their imperiousness, the more tyrannical are they towards it. This has been their uniform treatment towards us, and this character pervades all ranks of society, whether in public or private life.

"The only way to please John Bull is to give him a good beating, and, such is the singularity of his character that, the more you beat him, the greater is his respect for you, and the more he will esteem you. . . .

"If, after all I have now written, you still think that this war is unjust, and think it worth the trouble in order to ascertain the truth, I wish papa would take a trip across the Atlantic. If he is not convinced of the truth of what I have written in less than two months, I will agree to support myself all the time I am in England after this date, and never be a farthing's more expense to you. . . . I was glad to hear that Cousin Samuel Breese is in the navy. I really envy him very much. I hope one day, as a painter, I may be able to hand him down to posterity as an American Nelson. . . . As to my letters of introduction, I find that a painter and a visitor cannot be united. Were I to deliver my letters the acquaintance could not be kept up, and the bare thought of encountering the English reserve is enough to deter any one. . . . This objection, however, might be got over did it not take up so much time. Every moment is precious to me now. I don't know how soon I may be obliged to return home for want of means to support me; for the difficulties which are increasing in this country take off the attention of the people from the fine arts, and they withhold that patronage from young artists which they would, from their liberality, in other circumstances freely bestow. . . .

"You mention that some of the Ralston family are in Boston on a visit, and that Mr. Codman is attached to Eliza. Once in my life, you know, if you had told me this and I had been a very bloody-minded young man, who knows but Mr. Codman might have been challenged. But I suppose he takes advantage of my being in England. If it is as you say, I am very happy to hear it, for Elizabeth is a girl whom I very much esteem, and there is no doubt that she will make an excellent wife."

In a letter from his mother of July 6, 1813, she thus

reassures him: "Mr. Codman is married. He married a Miss Wheeler, of Newburyport, so you will have no need of challenging him on account of Eliza Ralston."

In a postscript to the letter of November 1, Morse adds: —

"I have just read the political parts of this letter to my good friend Mr. A——n, and he not only approves of the sentiments in it, but pays me a compliment by saying that I have expressed the truth and nothing but the truth in a very clear and proper manner, and hopes it may do good."

Among young Morse's friends in England at that time was Henry Thornton, philanthropist and member of Parliament. In a letter to his parents of January 1, 1813, he says: —

"Last Thursday week I received a very polite invitation from Henry Thornton, Esq., to dine with him, which I accepted. I had no introduction to him, but, hearing that your son was in the country, he found me out and has shown me every attention. He is a very pleasant, sensible man, but his character is too well known to you to need any eulogium from me.

"At his table was a son of Mr. Stephen, who was the author of the odious Orders in Council. Mr. Thornton asked me at table if I thought that, if the Orders in Council had been repealed a month or two sooner, it would not have prevented the war. I told him I thought it would, at which he was much pleased, and, turning to Mr. Stephen, he said: 'Do you hear that, Mr. Stephen? I always told you so.'

"Last Wednesday I dined at Mr. Wilberforce's. I was extremely pleased with him. At his house I met Mr.

Grant and Mr. Thornton, members of Parliament. In the course of conversation they introduced America, and Mr. Wilberforce regretted the war extremely; he said it was like two of the same family quarrelling; that he thought it a judgment on this country for its wickedness, and that they had been justly punished for their arrogance and insolence at sea, as well as the Americans for their vaunting on land.

"As Mr. Thornton was going he invited me to spend a day or two at his seat at Clapham, a few miles out of town. I accordingly went and was very civilly treated. The *reserve* which I mentioned in a former letter was evident, however, here, and I felt a degree of embarrassment arising from it which I never felt in America. The second day I was a little more at my ease.

"At dinner were the two sons of the Mr. Grant I mentioned above. They are, perhaps, the most promising young men in the country, and you may possibly one day hear of them as at the head of the nation. [One of these young men was afterwards raised to the peerage as Lord Glenelg.]

"After dinner I got into conversation with them and with Mr. Thornton, when America again became the topic. They asked me a great many questions respecting America which I answered to the best of my ability. They at length asked me if I did not think that the ruling party in America was very much under French influence. I replied 'No'; that I believed on the contrary that nine tenths of the American people were prepossessed strongly in favor of this country. As a proof I urged the universal prevalence of English fashions in preference to French, and English manners and customs; the universal rejoicings on the success of the English over the French; the marked attention shown to English travellers and visitors; the neglect with which they treated their own literary productions on account of the strong prejudice in favor of English works; that everything, in short, was enhanced in its value by having attached to it the name English.

"On the other hand, I told them that the French were a people almost universally despised in America, and by at least one half hated. As in England, they were esteemed the common enemies of mankind; that French fashions were discountenanced and loathed; that a Frenchman was considered as a man always to be suspected; that young men were forbidden by their parents, in many instances, to associate with them, they considering their company and habits as tending to subvert their morals, and to render them frivolous and insincere. I added that in America as well as everywhere else there were bad men, men of no principles, whose consciences never stand in the way of their ambition or avarice; but that I firmly believed that, as a body, the American Congress was as pure from corruption and foreign influence as any body of men in the world. They were much pleased with what I told them, and acknowledged that America and American visitors generally had been treated with too much contempt and neglect.

"In the course of the day I asked Mr. Thornton what were the objects that the English Government had in view when they laid the Orders in Council. He told me in direct terms, '*the Universal monopoly of Commerce*'; that they had long desired an excuse for such measures as the Orders in Council, and that the French decrees were exactly what they wished, and the opportunity was seized with avidity the moment it was offered. They knew that the Orders in Council bore hard upon the Americans, but they considered that as merely *incidental.*

"To this I replied that, if such was the case as he represented it, what blame could be attached to the American Government for declaring war? He said that it was urged that America ought to have considered the circumstances of the case, and that Great Britain was fighting for the liberties of the world; that America was, in a great degree, interested in the decision of the contest, and that she ought to be content to suffer a little.

"I told him that England had no right whatever to infringe on the neutrality of America, or to expect because she (England) supposed herself to have justice on her side in the contest with France, that, of course, the Americans should think the same. The moment America declared this opinion her neutrality ceased. 'Besides,' said I, 'how can they have the face to make such a declaration when you just now said that their object was universal monopoly, and they longed for an excuse to adopt measures to that end?' I told him that it showed that all the noise about England's fighting for the liberties of mankind proved to be but a thirst, a selfish desire for *universal monopoly.*

"This he said seemed to be the case; he could not deny it. He was going on to observe something respecting the French decrees when we were interrupted, and I have not been able again to resume the conversation. I returned to town with him shortly after in his carriage, where, as there were strangers, I could not introduce it again."

After this follow two long pages giving further reasons for the stand he has taken, which I shall not include, only quoting the following sentences towards the end of the letter: —

"You will have heard before this arrives of the glorious news from Russia. Bonaparte is for once *defeated*, and will probably never again recover from it.

"My regards to Mr. Stuart [Gilbert Stuart]. I feel quite flattered at his remembrance of me. Tell him that, by coming to England, I know how more justly to appreciate his great merits. There is really no one in England who equals him.

"Accompanying this are some newspapers, some of Cobbett's, a man of no principle and a great rascal, yet a man of sense and says many good things."

I have quoted at length from this letter in order that we may gain a clearer insight into the character of the man. While in no wise neglecting his main objects in life, he yet could not help taking a deep interest in public affairs. He was frank and outspoken in his opinions, but courteous withal. He abhorred hypocrisy and vice and was unsparing in his condemnation of both. He enjoyed a controversy and was quick to discover the weak points in his opponent's arguments and to make the most of them.

These characteristics he carried with him through life, becoming, however, broader-minded and more tolerant as he grew in years and experience.

Morse's father had given him many letters of introduction to eminent men in England. Most of these he neglected to deliver, pleading in extenuation of his apparent carelessness that he could not spare the time from his artistic studies to fulfill all the duties that would be expected of him in society, and that he also could not afford the expenses necessary to a well-dressed man.

The following note from William Wilberforce explains itself, but there seems to be some confusion of dates, for Morse had just said in his letter of January 1st that he dined at Mr. Wilberforce's over a week before.

KENSINGTON GORE,
January 4, 1813.

SIR, — I cannot help entertaining some apprehension of my not having received some letter or some card which you may have done me the favor of leaving at my house. Be this, however, as it may, I gladly avail myself of the sanction of a letter from your father for introducing myself to you; and, as many calls are mere matters of form, I take the liberty of begging the favor of your company at dinner on Wednesday next, at a quarter before five o'clock, at Kensington Gore (one mile from Hyde Park corner), and of thereby securing the pleasure of an acquaintance with you.

The high respect which I have always entertained for your father, in addition to the many obliging marks of attention which I have received from him, render me desirous of becoming personally known to you, and enable me with truth to assure you I am, with good will, sir,

Your faithful servant,
W. WILBERFORCE.

Among Morse's friends in London during the period of his student years, were Coleridge, Rogers, Lamb, and others whose names are familiar ones in the literary world.

While the letters of those days give only hints of the delightful intercourse between these congenial souls, the recollection of them was enshrined in the memory of some of their contemporaries, and the following reminiscences, preserved by Mr. James Wynne and recorded by Mr. Prime in his biography, will be found interesting:—

"Coleridge, who was a visitor at the rooms of Leslie and Morse, frequently made his appearance under the influence of those fits of despondency to which he was subject. On these occasions, by a preconcerted plan, they often drew him from this state to one of brilliant imagination.

"'I was just wishing to see you,' said Morse on one of these occasions when Coleridge entered with a hesitating step, and replied to their frank salutations with a gloomy aspect and deep-drawn sighs. 'Leslie and myself have had a dispute about certain lines of beauty; which is right?' And then each argued with the other for a few moments until Coleridge became interested, and, rousing from his fit of despondency, spoke with an eloquence and depth of metaphysical reasoning on the subject far beyond the comprehension of his auditors. Their point, however, was gained, and Coleridge was again the eloquent, the profound, the gifted being which his remarkable productions show him to be.

"'On one occasion,' said Morse, 'I heard him improvise for half an hour in blank verse what he stated to be a strange dream, which was full of those wonderful creations that glitter like diamonds in his poetical productions.' 'All of which,' remarked I, 'is undoubtedly lost to the world.' 'Not all,' replied Mr. Morse, 'for I recognize in the "Ancient Mariner" some of the thoughts of that evening; but doubtless the greater part, which would have made the reputation of any other man, perished with the moment of inspiration, never again to be recalled.'

"When his tragedy of 'Remorse,' which had a run of twenty-one nights, was first brought out, Washington Allston, Charles King, Leslie, Lamb, Morse, and Coleridge went together to witness the performance. They occupied a box near the stage, and each of the party was as much interested in its success as Coleridge himself.

"The effect of the frequent applause upon Coleridge was very manifest, but when, at the end of the piece, he was called for by the audience, the intensity of his emotions was such as none but one gifted with the fine sensibilities of a poet could experience. Fortunately the audience was satisfied with a mere presentation of himself. His emotions would have precluded the idea of his speaking on such an occasion.

"Allston soon after this became so much out of health that he thought a change of air and a short residence in the country might relieve him. He accordingly set out on his journey accompanied by Leslie and Morse.

"When he reached Salt Hill, near Oxford, he became so ill as to be unable to proceed, and requested Morse to return to town for his medical attendant, Dr. Tuthill, and Coleridge, to whom he was ardently attached.

"Morse accordingly returned, and, procuring a post-chaise, immediately set out for Salt Hill, a distance of twenty-two miles, accompanied by Coleridge and Dr. Tuthill.

"They arrived late in the evening and were busied with Allston until midnight, when he became easier, and Morse and Coleridge left him for the night.

"Upon repairing to the sitting-room of the hotel Morse opened Knickerbocker's 'History of New York,' which he had thrown into the carriage before leaving town. Coleridge asked him what work he had.

"'Oh,' replied he, 'it is only an American book.'

"'Let me see it,' said Coleridge.

"He accordingly handed it to him, and Coleridge was soon buried in its pages. Mr. Morse, overcome by the fatigues of the day, soon after retired to his chamber and fell asleep.

"On awakening next morning he repaired to the sitting-room, when what was his astonishment to find it still closed, with the lights burning, and Coleridge busy with the book he had lent him the previous night.

"'Why, Coleridge,' said he, approaching him, 'have you been reading the whole night?'

"'Why,' remarked Coleridge abstractedly, 'it is not late.'

"Morse replied by throwing open the blinds and permitting the broad daylight, for it was now ten o'clock, to stream in upon them.

"'Indeed,' said Coleridge, 'I had no conception of this; but the work has pleased me exceedingly. It is admirably written; pray, who is its author?'

"He was informed that it was the production of

Washington Irving. It is needless to say that, during the long residence of Irving in London, they became warm friends.

"At this period Mr. Abernethy was in the full tide of his popularity as a surgeon, and Allston, who had for some little time had a grumbling pain in his thigh, proposed to Morse to accompany him to the house of the distinguished surgeon to consult him on the cause of the ailment.

"As Allston had his hand on the bell-pull, the door was opened and a visitor passed out, immediately followed by a coarse-looking person with a large, shaggy head of hair, whom Allston at once took for a domestic. He accordingly enquired if Mr. Abernethy was in.

"'What do you want of Mr. Abernethy?' demanded this uncouth-looking person with the harshest possible Scotch accent.

"'I wished to see him,' gently replied Allston, somewhat shocked by the coarseness of his reception. 'Is he at home?'

"'Come in, come in, mon,' said the same uncouth personage.

"'But he may be engaged,' responded Allston. 'Perhaps I had better call another time.'

"'Come in, mon, I say,' replied the person addressed; and, partly by persuasion and partly by force, Allston, followed by Morse, was induced to enter the hall, which they had no sooner done than the person who admitted them closed the street door, and, placing his back against it, said: —

"'Now, tell me what is your business with Mr. Abernethy. I am Mr. Abernethy.'

"'I have come to consult you,' replied Allston, 'about an affection —'

"'What the de'il hae I to do with your affections?' bluntly interposed Abernethy.

"'Perhaps, Mr. Abernethy,' said Allston, by this time so completely overcome by the apparent rudeness of the eminent surgeon as to regret calling on him at all, 'you are engaged at present, and I had better call again.'

"'De'il the bit, de'il the bit, mon,' said Abernethy. 'Come in, come in.' And he preceded them to his office, and examined his case, which proved to be a slight one, with such gentleness as almost to lead them to doubt whether Abernethy within his consulting-room, and Abernethy whom they had encountered in the passage, was really the same personage."

While Morse was enjoying all these new experiences in England, the good people at home were jogging along in their accustomed ruts, but were deeply interested in the doings of the absent son and brother.

His mother writes on January 11, 1813: —

"Your letters are read with great pleasure by your acquaintance. I do not show those in which you say anything on *politics*, as I do not approve your *change*, and think it would only prejudice others. For that reason I do not wish you to write on that subject, as I love to read all your observations to your friends.

"We cannot get Edwards to be a ladies' man at all. He will not visit among the young ladies; he is as old as fifty, at least."

This same youthful misogynist and philosopher also writes to his brother on January 11: "I intend soon writing another letter in which I shall prove to your satisfaction that poetry is much superior to painting. You asserted the contrary in one of your letters, and brought an argument to prove it. I shall show the fallacy of that argument, and bring those to support my doctrine which are incontrovertible."

A letter from his friend, Mrs. Jarvis, the sister of his erstwhile flame, Miss Jannette Hart, informs him of the marriage of another sister to Captain Hull of the navy, commander of the Constitution. In this letter, written on March 4, 1813, at Bloomingdale, New York City, Mrs. Jarvis says: —

"I am in general proud of the spirit of my countrymen, but there is too little attention paid to the fine arts, to men of taste and science. Man here is weighed by his purse, not by his mind, and, according to the preponderance of that, he rises or sinks in the scale of individual opinion. A fine painting or marble statue is very rare in the houses of the rich of this city, and those individuals who would not pay fifty pounds for either, expend double that sum to vie with a neighbor in a piece of furniture.

"But do not tell tales. I would not say this to an Englishman, and I trust you have not yet become one. This, however, is poor encouragement for you to return to your native country. I hope better things of that country before you may return."

A friend in Philadelphia writes to him on May 3, 1813: —

"Your favor I received from the hands of Mr. King, and have been very much gratified with the introduction it afforded me to this worthy gentleman. You have doubtless heard of his safe arrival in our city, and of his having commenced his career in America, where, I am sorry to say, the arts are not, as yet, so much patronized as I hope to see them. Those of us who love them are too poor, and those who are wealthy regard them but little. I think, however, I have already witnessed an improvement in this respect, and the rich merchants and professional men are becoming more and more liberal in their patronage of genius, when they find it among native Americans.

"From the favorable circumstances under which your studies are progressing; from the unrivalled talents of the gentleman who conducts them; and, without flattery, suffer me to add, from the early proofs of your own genius, I anticipate, in common with many of our fellow citizens, the addition of one artist to our present roll whose name shall stand high among those of American painters.

"In your companion Leslie we also calculate on a very distinguished character.

"Our Academy of Fine Arts has begun the all-important study of the live figure. Mr. Sully, Mr. Peale, Mr. Fairman, Mr. King, and several others have devoted much attention to this branch of the school, and I hope to see it in their hands highly useful and improving.

"The last annual exhibition was very splendid *for us*. Some very capital landscapes were produced, many admirable portraits and one or two historical pictures.

"The most conspicuous paintings were Mr. Peale's picture of the 'Roman Charity' (or, if you please, the 'Grecian Daughter,' for Murphy has it so), and Mr. Sully's 'Lady of the Lake.'"

In a letter of May 30, 1813, to a friend, Morse says: —

"You ask in your letter what books I read and what I am painting. The little time that I can spare from painting I employ in reading and studying the old poets, Spenser, Chaucer, Dante, Tasso, etc. These are necessary to a painter.

"As to painting, I have just finished a large picture, eight feet by six feet six inches, the subject, the 'Death of Hercules,' which is now in the Royal Academy Exhibition at Somerset House. I have been flattered by the newspapers which seldom praise young artists, and they do me the honor to say that my picture, with that of another young man by the name of Monroe, form a distinguishing trait in this year's exhibition. . . .

"This praise I consider much exaggerated. Mr. West, however, who saw it as soon as I had finished it, paid me many compliments, and told me that, were I to live to his age, I should never make a better composition. This I consider but a compliment and as meant only to encourage me, and as such I receive it.

"I mention these circumstances merely to show that I am getting along as well as can be expected, and, if any credit attaches to me, I willingly resign it to my country, and feel happy that I can contribute a mite to her honor.

"The American character stands high in this country as to the production of artists, but in nothing else (except, indeed, I may now say *bravery*). Mr. West now stands at the head, and has stood ever since the arts began to flourish in this country, which is only about fifty years. Mr. Copley next, then Colonel Trumbull. Stuart in America has no rival here. As these are now old men and going off the stage, Mr. Allston succeeds in the prime of life, and will, in the opinion of the greatest connoisseurs in this country, carry the art to greater perfection than it ever has been carried either in ancient or modern times. . . . After him is a young man from Philadelphia by the name of Leslie, who is my room-mate."

How fallible is contemporary judgment on the claims of so-called genius to immortality. "For many are called, but few are chosen."

In another letter to his parents written about this time, after telling of his economies in order to make the money, advanced so cheerfully but at the cost of so much self-sacrifice on their part, last as long as possible, he adds:

"My greatest expense, next to *living*, is for canvas, frames, colors, etc., and visiting galleries. The frame of my large picture, which I have just finished, cost nearly twenty pounds, besides the canvas and colors, which cost nearly eight pounds more, and the frame was the cheapest I could possibly get. Mr. Allston's frame cost him sixty guineas.

"Frames are very expensive things, and, on that account, I shall not attempt another large picture for some time, although Mr. West advises me to paint *large* as much as possible.

"The picture which I have finished is 'The Death of Hercules'; the size is eight feet by six feet six inches. This picture I showed to Mr. West a few weeks ago, and he was extremely pleased with it and paid me very many high compliments; but as praise comes better from another than from one's self, I shall send you a complimentary note which Mr. West has promised to send me on the occasion.

"I sent the picture to the Exhibition at Somerset House which opens on the 3d of May, and have the satisfaction not only of having it received, but of having the praises of the council who decide on the admission of pictures. Six hundred were refused admission this year, so you may suppose that a picture (of the size of mine, too) must possess some merit to be received in preference to six hundred. A small picture may be received even if it is not very good, because it will serve to fill up some little space which would otherwise be empty, but a large one, from its excluding many smaller ones, must possess a great deal in its favor in order to be received.

"If you recollect I told you I had completed a model of a single figure of the same subject. This I sent to the Society of Arts at the Adelphi, to stand for the prize (which is offered every year for the best performance in painting, sculpture, and architecture and is a *gold medal*).

"Yesterday I received the note accompanying this, by which you will see that it is adjudged to me in sculpture this year. It will be delivered to me in public on the 13th of May or June, I don't know which, but I shall give you a particular account of the whole process as soon as I have received it. . . . I cannot close this letter without telling you how much I am indebted to that excellent man Mr. Allston. He is extremely partial to me and has often told me that he is proud of calling me his pupil. He visits me every evening and our conversation is generally upon the inexhaustible subject of our *divine* art, and upon *home* which is next in our thoughts.

"I know not in what terms to speak of Mr. Allston. I can truly say I do not know the slightest imperfection in him. He is amiable, affectionate, learned, possessed of the greatest powers of mind and genius, modest, unassuming, and, above all, a religious man. . . . I could write a quire of paper in his praise, but all I could say of him would give you but a very imperfect idea of him. . . .

"You must recollect, when you tell friends that I am studying in England, that I am a pupil of Allston and not Mr. West. They will not long ask who Mr. Allston is; he will very soon astonish the world. He claims me as his pupil, and told me a day or two since, in a jocose manner, that he should have a battle with Mr. West unless he gave up all pretension to me."

We gain further information concerning Morse's first triumphs, his painting and his statuette from the following reminiscences of a friend, Mr. Dunlap: —

"It was about the year 1812 that Allston commenced his celebrated picture of the 'Dead Man restored to Life by touching the Bones of Elisha,' which is now in the Pennsylvania Academy of Arts. In the study of this picture he made a model in clay of the head of the dead man to assist him in painting the expression. This was the practice of the most eminent old masters. Morse had begun a large picture to come out before the British public at the Royal Academy Exhibition. The subject was the 'Dying Hercules,' and, in order to paint it with the more effect, he followed the example of Allston and determined to model the figure in clay. It was his first attempt at modelling.

"His original intention was simply to complete such parts of the figure as were useful in the single view necessary for the purpose of painting; but, having done this, he was encouraged, by the approbation of Allston and other artists, to finish the entire figure.

"After completing it, he had it cast in plaster of Paris and carried it to show to West, who seemed more than pleased with it. After surveying it all round critically, with many exclamations of surprise, he sent his servant to call his son Raphael. As soon as Raphael made his appearance West pointed to the figure and said: 'Look there, sir; I have always told you any painter can make a sculptor.'

"From this model Morse painted his picture of the 'Dying Hercules,' of colossal size, and sent it, in May, 1813, to the Royal Academy Exhibition at Somerset House."

The picture was well received. A critic of one of the journals of that day in speaking of the Royal Academy thus notices Morse: —

"Of the academicians two or three have distinguished themselves in a preëminent degree; besides, few have added much to their fame, perhaps they have hardly sustained it. But the great feature in this exhibition is that it presents several works of very high merit by artists with whose performances, and even with whose names, we were hitherto unacquainted. At the head of this class are Messrs. Monroe and Morse. The prize of history may be contended for by Mr. Northcote and Mr. Stothard. We should award it to the former. After these gentlemen Messrs. Hilton, Turner, Lane, Monroe, and Morse follow in the same class." (London "Globe," May 14, 1813.)

In commemorating the "preëminent works of this exhibition," out of nearly two thousand pictures, this

THE DYING HERCULES

Painted by Morse in 1813

critic places the "Dying Hercules" among the first twelve.

On June 13, 1813, Morse thus writes to his parents: —

"I send by this opportunity (Mr. Elisha Goddard) the little cast of the Hercules which obtained the prize this year at the Adelphi, and also the gold medal, which was the premium presented to me, before a large assembly of the nobility and gentry of the country, by the Duke of Norfolk, who also paid me a handsome compliment at the same time.

"There were present Lord Percy, the Margravine of Anspach, the Turkish, Sardinian, and Russian Ambassadors, who were pointed out to me, and many noblemen whom I do not now recollect.

"My great picture also has not only been received at the Royal Academy, but has one of the finest places in the rooms. It has been spoken of in the papers, which you must know is considered a great compliment; for a young artist, unless extraordinary, is seldom or never mentioned till he has exhibited several times. They not only praise me, but place my picture among the most attractive in the exhibition. This I know will give you pleasure."

CHAPTER VI

JULY 10, 1813—APRIL 6, 1814

Letter from the father on economies and political views. — Morse deprecates lack of spirit in New England and rejoices at Wellington's victories. — Allston's poems. — Morse coat-of-arms. — Letter of Joseph Hillhouse. — Letter of exhortation from his mother. — Morse wishes to stay longer in Europe. — Amused at mother's political views. — The father sends more money for a longer stay. — Sidney exalts poetry above painting. — His mother warns him against infidels and actors. — Bristol. — Optimism. — Letter on infidels and his own religious observances. — Future of American art. — He is in good health, but thin. — Letter from Mr. Visger. — Benjamin Burritt, American prisoner. — Efforts in his behalf unsuccessful. — Capture of Paris by the Allies. — Again expresses gratitude to parents. — Writes a play for Charles Mathews. — Not produced.

THE detailed accounts of his economies which the young man sent home to his parents seem to have deeply touched them, for on July 10, 1813, his father writes to him: "Your economy, industry, and success in pursuing your professional studies give your affectionate parents the highest gratification and reward. We wish you to avoid carrying your economy to an *extreme*. Let your appearance be suited to the respectable company you keep, and your living such as will conduce most effectually to preserve health of body and vigor of mind. We shall all be willing to make sacrifices at home so far as may be necessary to the above purposes."

Farther on in this same letter the father says: "The character you give of Mr. Allston is, indeed, an exalted one, and we believe it correctly drawn. Your ardor has given it a high coloring, but the excess is that of an affectionate and grateful heart."

Referring to his son's political views, he answers in these broad-minded words: —

"I approve your love of your country and concern for its honor. Your errors, as we think them, appear to be the errors of a fair and honest mind, and are of a kind to be effectually cured by correct information of facts on both sides.

"Probably *we* may err because we are ignorant of many things which have fallen under your notice. We shall no doubt agree when we shall have opportunity to compare notes, and each is made acquainted with all that the other knows. I confidently expect an honorable peace in the course of six months, but may be deceived, as the future course of things cannot be foreseen.

"The present is one of the finest and most promising seasons I ever knew; the harvest to appearance will be very abundant. Heaven appears to be rewarding this part of the country for their conduct in opposing the present war."

Perhaps the good father did not mean to be malicious, but this is rather a wicked little thrust at the son's vehemently expressed political views. On this very same date, July 10, 1813, Morse writes to his parents: —

"I have just heard of the unfortunate capture of the Chesapeake. Is our infant Hercules to be strangled at his birth? Where is the spirit of former times which kindled in the hearts of the Bostonians? Will they still be unmoved, or must they learn from more bitter experience that Britain is not for peace, and that the only way to procure it is to join heart and hand in a vigorous prosecution of the war?

"It is not the time now to think of party; the country is in danger; but I hope to hear soon that the honor of our navy is retrieved. The brave Captain Lawrence will never, I am sure, be forgotten; his career of glory has been short but brilliant.

"All is rejoicing here; illuminations and fireworks and *feux de joie* for the capture of the Chesapeake and a victory in Spain.

"Imagine yourself, if possible, in my situation in an enemy's country and hearing songs of triumph and exultation on the misfortunes of my countrymen, and this, too, on the 4th of July. A less ardent spirit than mine might perhaps tolerate it, but I cannot. I do long to be at home, to be in the navy, and teach these insolent Englishmen how to respect us. . . .

"The Marquis Wellington has achieved a great victory in Spain, and bids fair to drive the French out very soon. At this I rejoice as ought every man who abhors tyranny and loves liberty. I wish the British success against everything but *my country*. I often say with Cowper: 'England, with all thy faults, I love thee still.'

"I am longing for Edwards' comparison between poetry and painting, and to know how he will prove the former superior to the latter. A painter *must* be a poet, but a poet need not be a painter. How will he get over this argument?

"By the way, Mr. Allston has just published a volume of poems, a copy of which I will endeavor to send you. They are but just published, so that the opinion of the public is not yet ascertained, but there is no doubt they will forever put at rest the calumny that America has never produced a poet.

"I have lately been enquiring for the coat-of-arms which belongs to the Morse family. For this purpose

I wish to know from what part of this Kingdom the Morses emigrated, and if you can recollect anything that belongs to the arms. If you will answer these questions minutely, I can, for half a crown, ascertain the arms and crest which belong to the family, which (as there is a degree of importance attached to heraldry in this country) may be well to know. I have seen the arms of one Morse which have been in the family three hundred years. So we can trace our antiquity as far as any family."

A letter from a college-mate, Mr. Joseph Hillhouse, written in Boston on July 12, 1813, gives a pretty picture of Morse's home, and contains some quaint gossip which I shall transcribe: —

"On Saturday afternoon the beauty of the weather invited my cousin Catherine Borland, my sister Mary (who is here on a visit), and myself to take a walk over to Charlestown for the purpose of paying a visit to your good parents. We found them just preparing tea, and at once concluded to join the family party.

"Present to the eye of your fancy the closing-in of a fine, blue-skied, sunny American Saturday evening, whose tranquillity and repose rendered it the fit precursor of the Sabbath. Imagine the tea-table placed in your sitting-parlor, all the windows open, and round it, first, the housekeeper pouring out tea; next her, Miss C. Borland; next her, your mother, whose looks spoke love as often as you were mentioned, and that was not infrequently, I assure you. On your mother's right sat my sister, next whom was your father in his long green-striped study gown, his apostolic smile responding to the eye of your mother when his dear son was his theme.

I was placed (and an honorable post I considered it) at his right hand.

"There the scene for you. Can you paint it? Neither of your brothers was at home. . . .

"In home news we have little variety. The sister of your quondam flame, Miss Ann Hart, bestowed her hand last winter on Victory as personified in our little fat captain, Isaac Hull, who is now reposing in the shade of his laurels, and amusing himself in directing the construction of a seventy-four at Portsmouth. Where the fair excellence, Miss Jannette herself, is at present, I am unable to say. The sunshine of her eyes has not beamed upon me since I beheld you delightedly and gallantly figuring at her side at Daddy Value's ball, where I exhibited sundry feats of the same sort myself.

"By the way, Mons. V. is still in fiddling condition, and the immaculate Ann Jane Caroline Gibbs, Madame, has bestowed a subject on the state!!

"A fortnight since your friend Nancy Goodrich was married to William Ellsworth. Emily Webster is soon to plight her faith to his brother Henry. Miss Mary Ann Woolsey thinks of consummating the blessedness of a Mr. Scarborough before the expiration of the summer. He is a widower of thirty or thirty-five with one child, a little girl four or five years old.

"Thus, you see, my dear friend, all here seem to be setting their faces heavenward; all seem ambitious of repairing the ravages of war. . . .

"P.S. Oh! horrid mistake I made on the preceding page! Nancy and Emily, on my knees I deprecate your wrath!! I have substituted William for Henry and

Henry for William. No, Henry is Nancy's and William Emily's. They are twins, and I, forsooth, must make them changelings!"

In a letter of July 30, 1813, his mother thus exhorts him: —

"I hope, my dear son, your success in your profession will not have a tendency to make you vain, or embolden you to look down on any in your profession whom Providence may have been less favorable to in point of talents for this particular business; and that you will observe a modesty in the reception of premiums and praises on account of your talents, that shall show to those who bestow them that you are worthy of them in more senses than merely as an artist. It will likewise convince those who are less favored that you are far from exulting in their disappointments, — as I hope is truly the case, — and prevent that jealousy and envy that too often discovers itself in those of the same profession. . . .

"We exceedingly rejoice in all your success, and hope you will persevere. Remember, my son, it is easier to get a reputation than to keep it unspotted in the midst of so much pollution as we are surrounded by. . . .

"C. Dexter thanks you for your attention to her request as it respects Southey's likeness. She does not wish you to take too much pains and trouble to get it, but she, I know, would be greatly pleased if you should send her one of him. If you should get acquainted with him, inform him that a very sensible, fine young lady in America requested it (but don't tell him her name) from having read his works."

In a long letter of August 10 and 26, 1813, after again

giving free rein to his political feelings, he returns to the subject of his art: —

"Mr. West promised me a note to you, but he is an old man and very forgetful, and I suppose he has forgotten it. I don't wish to remind him of it directly, but, if in the course of conversation I can contrive to mention it, I will. . . .

"With respect to returning home next summer, Mr. Allston and Mr. West think it would be an injury to me. Mr. Allston says I ought not to return till I am a *painter*. I long to return as much as you can wish to have me, but, if you can spare me a little longer, I should wish it. I abide your decision, however, completely. Mr. Allston will write you fully on this subject, and I will endeavor to persuade Mr. West also to do it.

"France I could not, at present, visit with advantage; that is to say for, perhaps, a year. Mr. Allston thinks I ought to be previously well grounded in the principles of the English school to resist the corruptions of the French school; for they are corrupt in the principles of painting, as in religion and everything else; but, when well grounded in the good principles of this school, I could study and select the few beauties of the French without being in danger of following their many errors. The Louvre also would, in about a year, be of the greatest advantage to me, and also the fine works in Italy. . . .

"Mama has amused me very much in her letter where she writes on politics. She says that, next to changing one's religion, she would dislike a man for changing his politics. Mama, perhaps, is not aware that she would in this way shut the door completely to conviction in anything. It would imply that, because a man is educated in error, he must forever live in error. I know exactly how mama feels; she thinks, as I did when at home, that it was impossible for the Federalists to be in the wrong; but, as all men are fallible, I think they may stand a chance of being wrong as well as any other class of people. . . .

"Mama thinks my '*error*' arises from wrong information. I will ask mama which of us is likely to get at the truth; I, who am in England and can see and hear all their motives for acting as they have done; or mama, who gets her information from the Federal papers, second-hand, with numerous additions and improvements made to answer party purposes, distorted and misrepresented?

"But to give you an instance. In the Massachusetts remonstrance they attribute the repeal of the Orders in Council to the kind disposition of the English Government, and a wish on their part to do justice, whereas it is notorious in this country that they repealed them on account of the injury it was doing themselves, and took America into consideration about as much as they did the inhabitants of Kamschatka. The conditional repeal of the Berlin and Milan decrees was a back door for them, and they availed themselves of it to sneak out of it. This necessity, this act of dire necessity, the Federal papers cry up as evincing a most forbearing spirit towards us, and really astonish the English themselves who never dreamt that it could be twisted in that way.

"Mama assigns as a reason for my thinking well of the English that they have been very polite to me, and that it is ingratitude in me if I do otherwise. A few individuals have treated me politely, and I do feel thankful and gratified for it; but a little politeness from an individual of one nation to an individual of another is certainly not a reason that the former's Government should be esteemed incapable of wrong by the latter. I esteem the English as a nation; I rejoice in their conquests on the Continent, and would love them heartily, if they would let me; but I am afraid to tell them this, they are already too proud.

"Their treatment of America is the worse for it. They are like a poor man who has got a lottery ticket and draws a great prize, and when his poor neighbor comes sincerely to congratulate him on his success, he holds up his head, and, turning up his nose, tells him that now he is his superior and then kicks him out of doors.

"Papa says he expects peace in six months. It may be in the disposition of America to make peace, but not in the will of the English. It is in the power of the Federalists to force her to peace, but they will not do it, so she will force us to do it."

As in most discussions, political or otherwise, neither party seems to have been convinced by the arguments of the other, for the parents continue to urge him to leave politics alone; indeed, they insist on his doing so. They also urge him to make every effort to support himself, if he should decide to spend another year abroad, for they fear that they will be unable to send him any more money. However, the father, when he became convinced that it was really to his son's interest to spend another year abroad, contrived to send him another thousand dollars. This was done at the cost of great self-sacrifice on the part of himself and his family, and was all the more praiseworthy on that account.

In a letter from his brother Edwards, written also on the 17th of November, is this passage: "I must defer giving my reasons for thinking Poetry superior to Painting; I will mention only a few of the principles upon which I found my judgment. Genius in both these arts is the power of making impressions. The question then is: which is capable of making the strongest impression; which can impress upon the mind most strongly a sublime or a beautiful idea? Does the sublimest passage in Milton excite a stronger sensation in the mind of a man of taste than the sublimest painting of Michael Angelo? Or, to make the parallel more complete, does Michael Angelo convey to you a stronger impression of the Last Judgment, by his painting, than Milton could by his poetry? Could Michael Angelo convey a more sublime idea of Death by his painting than Milton has in his 'Paradise Lost'? These are the principles upon which your 'divine art' is to be degraded below Poetry."

This was rather acute reasoning for a boy of twenty who had spent his life in the Boston and New Haven of those early days. The fact that he had never seen a great painting, whereas he had greedily read the poets, will probably account for his strong partisanship.

The pious mother writes on November 25, 1813: —

"With regard to the Americans being despised and hated in England, you were apprised by your Uncle Salisbury and others before you left this country that that was the case, and you ought not to be surprised when you realized it. The reason given was that a large portion of those who visit Europe are *dissipated infidels*, which has justly given the English a bad opinion of us as a nation. But we are happy to find that there are many exceptions to these, who do honor to the country which gave them birth, such as a West, an Allston, and many others, among whom, I am happy to say, we hope that you, my son, will be enrolled at no very distant day. . . .

"You mention being acquainted with young Payne, the play actor. I would guard you against any acquaintance with that description of people, as it will, sooner or later, have a most corrupting effect on the morals, and, as a man is known by the company he keeps, I should be very sorry to have you enrolled with such society, however pure you may believe his morals to be.

"Your father and myself were eleven days in company with him in coming from Charleston, South Carolina. His behavior was quite unexceptionable then, but he is in a situation to ruin the best morals. I hope you do not attend the theatre, as I have ever considered it a most bewitching amusement, and ruinous both to soul and body. I would therefore guard you against it."

His brother Richard joined the rest of the family in urging the young and impulsive artist to leave politics alone, as we learn from the following words which begin a letter of November 27, 1813: —

MY DEAR BROTHER, — Your letters by the Neptune, and also the medal, gave us great pleasure. The politics, however, were very disagreeable and occupied no inconsiderable part of your letters. Your kind wishes for *our* reformation we must beg leave to retort by hoping for *your* speedy amendment.

There are gaps in the correspondence of this period. Many of the letters from both sides of the Atlantic seem never to have reached their destination, owing to the disturbed state of affairs arising from the war between the two countries.

The young artist had gone in October, 1813, to Bristol, at the earnest solicitation of friends in that city, and seems to have spent a pleasant and profitable five months there, painting a number of portraits. He refers to letters written from Bristol, but they were either never received or not preserved. Of other letters I have only fragments, and some that are quoted by Mr. Prime in his biography have vanished utterly. Still, from what remains, we can glean a fairly good idea of the life of the young man at that period. His parents continually begged him to leave politics alone and to tell them more of his artistic life, of his visits to interesting places, and of his intercourse with the literary and artistic celebrities of the day.

We, too, must regret that he did not write more fully on these subjects, for there must have been a mine of interesting material at his disposal. We also learn that there seems to have been a strange fatality attached to the little statuette of the "Dying Hercules," for, although he packed it carefully and sent it to Liverpool on June 13, 1813, to be forwarded to his parents, it never reached them until over two years later. The superstitious will say that the date of sending may have had something to do with this.

Up to this time everything, except the attitude of England towards America, had been *couleur de rose* to the enthusiastic young artist. He was making rapid progress in his studies and was receiving the encomiums of his fellow artists and of the critics. His parents were denying themselves in order to provide the means for his support, and, while he was duly appreciative of their goodness, he could not help taking it more or less as a matter of course. He was optimistic with regard to the future, falling into the common error of gifted young artists that, because of their artistic success, financial success must of necessity follow. He had yet to be proved in the school of adversity, and he had not long to wait. But I shall let the letters tell the story better than I can. The last letter from him to his parents from which I have quoted was written on August 12 and 26, 1813.

On March 12, 1814, he writes from London after his return from Bristol: —

"There is a great drawback to my writing long letters to you; I mean the uncertainty of their reaching you.

"Mama's long letter gave me particular pleasure. Some of her observations, however, made me smile, especially the reasons she assigns for the contempt and hatred of England for America. First, I am inclined to doubt the fact of there being so many *infidel* Americans in the country; second, if there were, there are not so many *religious* people here who would take the pains to enquire whether they had religion or not; and third, it is not by seeing the individual Americans that an opinion unfavorable to us is prevalent in England. . . .

"With respect to my religious sentiments, they are unshaken; their influence, I hope, will always guide me through life. I hear various preachings on Sundays, sometimes Mr. Burder, but most commonly the Church of England clergy, as a church is in my neighborhood and Mr. B.'s three miles distant. I most commonly heard Dr. Biddulph, of St. James's Church, a most excellent, orthodox, evangelical man. I was on the point many times of going to hear Mr. Lowell, who is one of the dissenting clergymen of Bristol, but, as the weather proved very unfavorable, uncommonly so every Sunday I was there, and I was at a great distance from his church, I was disappointed. I shall endeavor to hear him preach when I go back to Bristol again."

This was in reply to many long exhortations in his parents' letters, and especially in his mother's, couched in the extravagant language of the very pious of those days, to seek first the welfare of his "never-dying soul."

"I have returned from Bristol to attend the exhibitions and to endeavor to get a picture into Somerset House. My stay in Bristol was very pleasant, indeed, as well as profitable. I was there five months and, in May, shall probably go again and stay all summer. I was getting into good business in the portrait way there, and, if I return, shall be enabled, probably, to support myself as long as I stay in England.

"The attention shown me by Mr. Harman Visger and family, whom I have mentioned in a former letter, I shall never forget. He is a rich merchant, an American (cousin to Captain Visscher, my fellow passenger, by whom I was introduced to him). He has a family of seven children. I lived within a few doors of him, and was in and out of his house ever day. . . .

Four pages of this letter are, unfortunately, missing. It begins again abruptly:—

". . . prevented by illness from writing you before.

I shall endeavor to support myself, if not, necessity will compel me to return home an unfinished painter; it depends altogether on circumstances. I may get a good run of portraits or I may not; it depends so much on the whim of the public; if they should happen to fancy my pictures, I shall succeed; if not, why, I shall not succeed. I am, however, encouraged to hope. . . .

"If I am prohibited from writing or thinking of politics, I hope my brothers will not be so ungenerous as to give me any. . . .

"Mr. Allston's large picture is now exhibiting in the British Gallery. It has excited a great deal of curiosity and he has obtained a wonderful share of praise for it. . . . The picture is very deservedly ranked among the highest productions of art, either in ancient or modern times. It is really a pleasant consideration that the palm of painting still rests with America, and is, in all probability, destined to remain with us. All we wish is a taste in the country and a little more wealth. . . . In order to create a taste, however, pictures, first-rate pictures, must be introduced into the country, for taste is only acquired by a close study of the merits of the old masters. In Philadelphia I am happy to find they have successfully begun. I wish Americans would unite in the thing, throw aside local prejudices and give their support to *one* institution. Let it be in Philadelphia, since is is so happily begun there, and let every American feel a pride in supporting that institution; let it be a national not a city institution. Then might the arts be so encouraged that Americans might remain at home and not, as at present, be under the painful necessity of exiling themselves from their country and their friends.

"This will come to pass in the course of time, but not in my day, I fear, unless there is more exertion made to forward the arts than at present. . . ."

In this he proved a true prophet, and, as we shall see later, his exertions were a potent factor in establishing the fine arts on a firm basis in New York.

"I am in very good health and I hope I feel grateful for it. I have not been ill for two days together since I have been in England. I am, however, of the *walking-stick* order, and think I am thinner than I was at home. They all tell me so. I'm not so good-looking either, I am told; I have lost my color, grown more sallow, and have a face approaching to the hatchet class; but none of these things concern me; if I can paint good-looking, plump ladies and gentlemen, I shall feel satisfied. . . .

"We have had a dreadfully severe winter here in England, such as has not been known for twenty-two years. When I came from Bristol the snow was up on each side of the road as high as the top of the coach in many places, especially on Marlborough Down and Hounslow Heath."

His friend Mr. Visger thus writes to him from Bristol on April 1, 1814: —

"It gave me pleasure to learn that Mr. Leslie sold his picture of Saul, etc., at so good a price. I hope it will stimulate a friend of his to use his best exertions and time to endeavor even to excel the 'Witch of Endor.' I think I perceive a few symptoms of amendment in him, and the request of his father that he must support himself is, in the opinion of his friends here, the best thing that could have befallen him. He will now have the pleasure to taste the sweets of his own labor, and I hope will, in reality, know what true independence is. Let him not despair and he will certainly succeed.

"Excuse my having taken up so much of your time in reading what I have written about Mr. Leslie's friend; I hope it will not make the pencil work less smoothly.

"It gave us all great pleasure to hear that Mr. Allston's 'Dead and Alive Man' got the prize. It would be a great addition to our pleasure to hear that those encouragers of the fine arts have offered him fifteen hundred or two thousand guineas for it. . . .

"There is an old lady waiting your return to have her portrait painted. Bangley says one or two more are enquiring for Mr. Morse.

"You seem to have forgotten your friend in Stapleton prison. Did you not succeed in obtaining his release?"

This refers to a certain Mr. Benjamin Burritt, an American prisoner of war. Morse used every effort, through his friend Henry Thornton, to secure the release of Mr. Burritt. On December 30, 1813, he wrote to Mr. Thornton from Bristol: —

RESPECTED SIR, — I take the liberty of addressing you in behalf of an American prisoner of war now in the Stapleton depot, and I address you, sir, under the conviction that a petition in the cause of humanity will not be considered by you as obtrusive.

The prisoner I allude to is a gentleman of the name of Burritt, a native of New Haven, in the State of Connecticut; his connections are of the highest respectability in that city, which is notorious for its adherence to Federal principles. His friends and relatives are among my father's friends, and, although I was not, until now, personally acquainted with him, yet his face is familiar to me, and many of his relatives were my particular friends while I was receiving my education at Yale College in New Haven. From that college he was graduated in the year ——. A classmate of his was the Reverend Mr. Stuart, who is one of the professors of the Andover Theological Institution, and of whom, I think, my father has spoken in some of his letters to Mr. Wilberforce.

Mr. Burritt, after he left college, applied himself to study, so much so as to injure his health, and, by the advice of his physicians, he took to the sea as the only remedy left for him. This had the desired effect, and he was restored to health in a considerable degree.

Upon the breaking out of the war with this country, all the American coasting trade being destroyed, he took a situation as second mate in the schooner Revenge, bound to France, and was captured on the 10th of May, 1813.

Since that time he has been a prisoner, and, from the enclosed certificates, you will ascertain what has been his conduct. He is a man of excellent religious principles, and, I firmly believe, of the strictest integrity. So well assured am I of this that, in case it should be required, *I will hold myself bound to answer for him in my own person.*

His health is suffering by his confinement, and the unprincipled society, which he is obliged to endure, is peculiarly disagreeable to a man of his education.

My object in stating these particulars to you, sir, is (if possible and consistent with the laws of the country), to obtain for him, through your influence, his liberty on his parole of honor. By so doing you will probably be the means of preserving the life of a good man, and will lay his friends, my father, and myself under the greatest obligations.

Trusting to your goodness to pardon this intrusion upon your time, I am, sir, with the highest consideration,

Your most humble, obedient servant,

SAMUEL F. B. MORSE.

To this Mr. Thornton replied: —

DEAR SIR, — You will perceive by the enclosed that there is, unhappily, no prospect of our effecting our wishes in respect to your poor friend at Bristol. I shall be glad to know whether you have had any success in obtaining a passport for Dr. Cushing.

I am, dear sir, yours, etc.

H. THORNTON.

The enclosure referred to by Mr. Thornton was the following letter addressed to him by Lord Melville: —

SIR, — Mr. Hay having communicated to me a letter which he received from you on the subject of Benjamin Burritt, an American prisoner of war in the depot at Stapleton, I regret much that, after consulting on this case with Sir Rupert George, and ascertaining the usual course of procedure in similar instances, I cannot discover any circumstances that would justify a departure from the rules observed toward other prisoners of the same description.

There can be no question that his case is a hard one, but I am afraid that it is inseparable from a state of war. It is not only not a solitary instance among the French and American prisoners, but, unless we were prepared to adopt the system of releasing all others of the same description, we should find that the number who might justly complain of undue partiality to this man would be very considerable.

I have the honor to be, sir, your most obedient and very humble servant,

MELVILLE.

This was a great disappointment to Morse, who had set his heart on being the means of securing the liberty of this unfortunate man. He was compelled to bow to the inevitable, however, and after this he did what he could to make the unhappy situation of the prisoner more bearable by extending to him financial assistance, although he had but little to spare at that time himself, and could but ill afford the luxury of giving.

Great events were occurring on the Continent at this time, and it is interesting to note how the intelligence of them was received in England by an enthusiastic student, not only of the fine arts, but of the humanities, who felt that, in this case, his sympathies and those of his family were in accord: —

April 6, 1814.

MY DEAR PARENTS, — I write in much haste, but it is to inform you of a most glorious event, no less than the capture of Paris, by the Allies. They entered it last Thursday, and you may conceive the sensations of the people of England on the occasion. As the cartel is the first vessel which will arrive in America to carry the news, I hope I shall have the great satisfaction of hearing that

I am the first who shall inform you of this great event; the particulars you will see nearly as soon as this.

I congratulate you and the rest of the good people of the world on the occasion. *Despotism* and *Usurpation* are fallen, never, I hope, to rise again. But what gives me the greatest pleasure in the contemplation of this occurrence is the spirit of religion and, consequently, of humanity which has constantly marked the conduct of the Allies. Their moderation through all their unparallelled successes cannot be too much extolled; they merit the grateful remembrance of posterity, who will bless them as the restorers of a blessing but little enjoyed by the greater part of mankind for centuries. I mean the inestimable blessing of *Peace*.

But I must cut short my feelings on the subject; were I to give them scope they would fill quires; they are as ardent as yours possibly can be. Suffice it to say that I see the hand of Providence so strongly in it that I think an infidel must be converted by it, and I hope I feel as a Christian should on such an occasion.

I am well, in excellent spirits and shall use my utmost endeavors to support myself, for now more than ever is it necessary for me to stay in Europe. Peace is inevitable, and the easy access to the Continent and the fine works of art there render it doubly important that I should improve them to my utmost.

I cannot ask more of my parents than they have done for me, but the struggle will be hard for me to get along and improve myself at the same time. Portraits are the only things which can support me at present, but it is insipid, indeed, for one who wishes to be at the head of the first branch of the art, to be stopped halfway, and be obliged to struggle with the difficulty of maintaining himself, in addition to the other difficulties attendant on the profession.

But it is impossible to place this in a clear light in a letter. I wish I could talk with you on the subject, and I could in a short time make it clear to you. I cannot ask it of you and I do not till I try what I can do. You have already done more than I deserved and it would be ingratitude in me to request more of you, and I do not; only I say these things that you may not expect so much from me in the way of improvement as you may have been led to suppose.

Morse seems to have made an excursion into dramatic literature at about this time, as the following draft of a letter, without date, but evidently written to the celebrated actor Charles Mathews, will testify: —

Not having the honor of a personal acquaintance with you, I have taken the liberty of enclosing to you a farce which, if, on perusal, you should think worthy of the stage, I beg you to accept, to be performed, if consistent with your plans, on the night appointed for your benefit.

If I should be so much favored as to obtain your good opinion of it, the approbation alone of Mr. Mathews will be a sufficient reward for the task of writing it.

The pleasure which I have so often received from you in the exercise of your comic powers would alone prompt me to make some return which might show you, at least, that I can be grateful to those who have at any time afforded me pleasure.

With respect to your accepting or not accepting it, I

wish you to act your pleasure entirely. If you think it will be of benefit to you by drawing a full house, or in any other way, it is perfectly at your service. If you think it will not succeed, will you have the goodness to enclose it under cover and direct to Mr. T. G. S., artist, 82 Great Titchfield Street; and I assure you beforehand that you need be under no apprehension of giving me mortification by refusing it. It would only convince me that I had not dramatic talents, and would serve, perhaps, to increase my ardor in the pursuit of my professional studies. If, however, it should meet with your approbation and you should wish to see me on the subject, a line directed as above enclosing your address shall receive immediate attention.

I am as yet undecided what shall be its name. The character of Oxyd I had designed for you. The farce is a first attempt and has received the approbation, not only of my theatrical friends generally, but of some confessed critics by whom it has been commended.

With sentiments of respect and esteem I remain,

Your most obedient humble servant,

T. G. S.

As no further mention of this play is made I fear that the great Charles Mathews did not find it available. There is also no trace of the play itself among the papers, which is rather to be regretted. We can only surmise that Morse came to the conclusion (very wisely) that he had no "dramatic talents," and that he turned to the pursuit of his professional studies with increased ardor.

CHAPTER VII

MAY 2, 1814—OCTOBER 11, 1814

Allston writes encouragingly to the parents.—Morse unwilling to be mere portrait-painter.—Ambitious to stand at the head of his profession.—Desires patronage from wealthy friends.—Delay in the mails.—Account of *entrée* of Louis XVIII into London.—The Prince Regent.—Indignation at acts of English.—His parents relieved at hearing from him after seven months' silence.—No hope of patronage from America.—His brothers.—Account of fêtes.—Emperor Alexander, King of Prussia, Blücher, Platoff.—Wishes to go to Paris.—Letter from M. Van Schaick about battle of Lake Erie.—Disgusted with England.

MORSE had now spent nearly three years in England. He was maturing rapidly in every way, and what his master thought of him is shown in this extract from a letter of Washington Allston to the anxious parent at home:—

"With regard to the progress which your son has made, I have the pleasure to say that it is unusually great for the time he has been studying, and indeed such as to make me proud of him as a pupil and to give every promise of future eminence. . . .

"Should he be obliged to return *now* to America, I much fear that all which he has acquired would be rendered abortive. It is true he could there paint very good portraits, but I should grieve to hear at any future period that, on the foundation now laid, he shall have been able to raise no higher superstructure than the fame of a portrait-painter. I do not intend here any disrespect to portrait-painting; I know it requires no common talent to excel in it. . . .

"In addition to this *professional report* I have the sincere satisfaction to give my testimony to his conduct as a man, which is such as to render him still worthy of being affectionately remembered by his moral and religious friends in America. This is saying a great deal for a young man of two-and-twenty in London, but is not more than justice requires me to say of him."

On May 2, 1814, Morse writes home: —

"You ask if you are to expect me the next summer. This leads me to a little enlargement on the peculiar circumstances in which I am now placed. Mr. Allston's letter by the same cartel will convince you that industry and application have not been wanting on my part, that I have made greater progress than young men generally, etc., etc., and of how great importance it is to me to remain in Europe for some time yet to come. Indeed I feel it so much so myself that I shall endeavor to stay at all risks. If I find that I cannot support myself, that I am contracting debts which I have no prospect of paying, I shall then return home and settle down into a mere portrait-painter for some time, till I can obtain sufficient to return to Europe again; for I cannot be happy unless I am pursuing the intellectual branch of the art. Portraits have none of it; landscape has some of it, but history has it wholly. I am certain you would not be satisfied to see me sit down quietly, spending my time in painting portraits, throwing away the talents which Heaven has given me for the higher branches of art, and devoting my time only to the inferior.

"I need not tell you what a difficult profession I have undertaken. It has difficulties in itself which are sufficient to deter any man who has not firmness enough to go through with it at all hazards, without meeting with any obstacles aside from it. The more I study it, the more I am enchanted with it; and the greater my progress, the more am I struck with its beauties, and the perseverance of those who have dared to pursue it through the thousands of natural hindrances with which the art abounds.

"I never can feel too grateful to my parents for having assisted me thus far in my profession. They have done more than I had any right to expect; they have conducted themselves with a liberality towards me, both in respect to money and to countenancing me in the pursuit of one of the noblest of professions, which has not many equals in this country. I cannot ask of them more; it would be ingratitude.

"I am now in the midst of my studies when the great works of ancient art are of the utmost service to me. Political events have just thrown open the whole Continent; the whole world will now leave war and bend their attention to the cultivation of the arts of peace. A golden age is in prospect, and art is probably destined to again revive as in the fifteenth century.

"The Americans at present stand unrivalled, and it is my great ambition (and it is certainly a commendable one) to stand among the first. My country has the most prominent place in my thoughts. How shall I raise her name, how can I be of service in refuting the calumny, so industriously spread against her, that she has produced no men of genius? It is this more than anything (aside from painting) that inspires me with a desire to excel in my art. It arouses my indignation and gives me tenfold energy in the pursuit of my studies. I should like to be the greatest painter *purely out of revenge.*

"But what a damper is thrown upon my enthusiasm

when I find that, the moment when all the treasures of art are before me, just within my reach; that advantages to the artist were never greater than now; Paris with all its splendid depository of the greatest works but a day or two's journey from me, and open to my free inspection,—what a damper, I say, is it to find that my three years' allowance is just expired; that while all my contemporary students and companions are revelling in these enjoyments, and rapidly advancing in their noble studies, they are leaving me behind, either to return to my country, or, by painting portraits in Bristol, just to be able to live through the year. The thought makes me melancholy, and, for the first time since I left home, have I had one of my desponding fits. I have got over it now, for I would not write to you in that mood for the world. My object in stating this is to request patronage from some rich individual or individuals for a year or two longer at the rate of £250 per year. This to be advanced to me, and, if required, to be returned in money as soon as I shall be able, or by pictures to the amount when I have completed my studies. . . . If Uncle Salisbury or Miss Russell could do it, it would be much more grateful to me than from any others. . . .

"The box containing my plaster cast I found, on enquiry, is still at Liverpool where it has been, to my great disappointment, now nearly a year. I have given orders to have it sent by the first opportunity. Mr. Wilder will tell you that he came near taking out my great picture of the Hercules to you. It seems as though it is destined that nothing of mine shall reach you. I packed it up at a moment's warning and sent it to Liverpool to go by the cartel, and I found it arrived the day after she had sailed. I hope it will not be long before both the boxes will have an opportunity of reaching you.

"I am exceedingly sorry you have forgotten a passage in one of my letters where I wished you not to feel anxious if you did not hear from me as often as you had done. I stated the reason, that opportunities were less frequent, more circuitous, and attended with greater interruptions. I told you that I should write at least once in three weeks, and that you must attribute it to anything but neglect on my part.

"Your last letter has hurt me considerably, for, owing to some accident or other, my letters have miscarried, and you upbraid me with neglect, and fear that I am not as industrious or correct as formerly. I know you don't wish to hurt me, but I cannot help feeling hurt when I think that my parents have not the confidence which I thought they had in me; that some interruptions, which all complain of and which are natural to a state of warfare, having prevented letters, which I have written, from being received; instead of making allowances for these things, to have them attribute it to a falling-off in industry and attention wounds me a great deal. Mrs. Allston, to her great surprise, received just such a letter from her friends, and it hurt her so that she was ill in consequence. . . .

"I dine at Mr. Macaulay's at five o'clock to-day, and shall attend the House of Commons to-morrow evening, where I expect to hear Mr. Wilberforce speak on the Slave Trade, with reference to the propriety of making the universal abolition of it an article in the pending negotiations. If I have time in this letter I will give you some account of it. In the mean time I will give you a slight account of some scenes of which I have been a happy witness in the great drama now acting in the Theatre of Europe.

"You will probably, before this reaches you, hear of the splendid *entrée* of Louis XVIII into London. I was a spectator of this scene. On the morning of the day, about ten o'clock, I went into Piccadilly through which the procession was to pass. I did not find any great concourse of people at that hour except before the Pultney Hotel, where the sister of Emperor Alexander resides on a visit to this country, the Grand Duchess of Oldenburg. I thought it probable that, as the procession would pass this place, there would be some uncommon occurrence taking place before it, so I took my situation directly opposite, determined, at any rate, to secure a good view of what happened.

"I waited four or five hours, during which time the people began to collect from all quarters; the carriages began to thicken, the windows and fronts of the houses began to be decorated with the white flag, white ribbons, and laurel. Temporary seats were fitted up on all sides, which began to be filled, and all seemed to be in preparation. About this time the King's splendid band of music made its appearance, consisting, I suppose, of more than fifty musicians, and, to my great gratification, placed themselves directly before the hotel. They began to play, and soon after the grand duchess, attended by several Russian noblemen, made her appearance on the balcony, followed by the Queen of England, the Princess Charlotte of Wales, the Princess Mary, Princess Elizabeth, and all the female part of the royal family. From this fortunate circumstance you will see that I had an excellent opportunity of observing their persons and countenances.

"The Duchess of Oldenburg is a common-sized woman of about four or five and twenty; she has rather a pleasant countenance, blue eyes, pale complexion, regular features, her cheek-bones high, but not disagreeably so. She resembles very much her brother the Emperor, judging from his portrait. She had with her her little nephew, Prince Alexander, a boy of about three or four years old. He was a lively little fellow, playing about, and was the principal object of the attention of the royal family.

"The Queen, if I was truly directed to her, is an old woman of very sallow complexion, and nothing agreeable either in her countenance or deportment; and, if she was not called a queen, she might as well be any ugly old woman. The Princess Charlotte of Wales I thought pretty; she has small features, regular, pale complexion, great amiability of expression and condescension of manners; the Princess Elizabeth is extremely corpulent, and, from what I could see of her face, was agreeable though nothing remarkable.

"One of the others, I think it was the Princess Mary, appeared to have considerable vivacity in her manners; she was without any covering to her head, her hair was sandy, which she wore cropped; her complexion was probably fair originally, but was rather red now; her features were agreeable.

"It now began to grow late, the people were beginning to be tired, wanting their dinners, and the crowd to thicken, when a universal commotion and murmur through the crowd and from the housetops indicated that the procession was at hand. This was followed by the thunder of artillery and the huzzas of the people toward the head of the street, where the houses seemed to be alive with the twirling of hats and shaking of handkerchiefs. This seemed to mark the progress of the King; for, as he came opposite each house, these actions became most violent, with cries of '*Vivent les Bourbons!*' '*Vive le Roi!*' '*Vive Louis!*' etc.

"I now grew several inches taller; I stretched my neck and opened my eyes. One carriage appeared, drawn by six horses, decorated with ribbons, and containing some of the French *noblesse;* another, of the same description, with some of the French royal family. At length came a carriage drawn by eight beautiful Arabian cream-colored horses. In this were seated Louis XVIII, King of France, the Prince Regent of England, the Duchesse d'Angoulême, daughter of Louis XVI, and the Prince of Condé. They passed rather quickly, so that I had but a glance at them, though a distinct one. The Prince Regent I had often seen before; the King of France I had a better sight of afterwards, as I will presently relate. The Duchesse d'Angoulême had a fine expression of countenance, owing probably to the occasion, but a melancholy cast was also visible through it; she was pale. The Prince of Condé I have no recollection of.

"After this part of the procession had passed, the crowd became exceedingly oppressive, rushing down the street to keep pace with the King's carriage. As the King passed the royal family he bowed, which they returned by kissing their hands to him and shaking their handkerchiefs with great enthusiasm. After they had gone by, the royal family left the balcony, where they had been between two and three hours.

"My only object now was to get clear of the crowd. I waited nearly three quarters of an hour, and at length, by main strength, worked myself edgewise across the street, where I pushed down through stables and houses and by-lanes to get thoroughly clear, not caring where I went, as I knew I could easily find my way when I got into a street. This I at last gained, and, to my no small astonishment, found myself by mere chance directly opposite the hotel where Louis and his suite were.

"The Prince Regent had just left the place, and with his carriage went a great part of the mob, which left the space before the house comparatively clear. It soon filled again; I took advantage, however, and got directly before the windows of the hotel, as I expected the King would show himself, for the people were calling for him very clamorously.

"I was not disappointed, for, in less than half a minute he came to the window, which was open, before which I was. I was so near him I could have touched him. He stayed nearly ten minutes, during which time I observed him carefully. He is very corpulent, a round face, dark eyes, prominent features; the character of countenance much like the portraits of the other Louises; a pleasant face, but, above all, such an expression of the moment as I shall never forget, and in vain attempt to describe.

"His eyes were suffused with tears, his mouth slightly open with an unaffected smile full of gratitude, and seemed to say to every one, 'Bless you.' His hands were a little extended sometimes as if in adoration to heaven, at others as if blessing the people. I entered into his feelings. I saw a monarch who, for five-and-twenty years, had been an exile from his country, deprived of his throne, and, until within a few months, not a shadow of a hope remaining of ever returning to it again. I saw him raised, as if by magic, from a private station in an instant to his throne, to reign over a nation which has made itself the most conspicuous of any nation on the globe. I tried to think as he did, and, in the heat of my enthusiasm, I joined with heart and soul in the cries of '*Vive le roi!*' '*Vive Louis!*' which rent the air from the mouths of thousands. As soon as he left the window, I returned home much fatigued, but well satisfied that my labor had not been for naught. . . .

"Mr. Wilberforce is an excellent man; his whole soul is bent on doing good to his fellow men. Not a moment of his time is lost. He is always planning some benevolent scheme or other, and not only planning but executing; he is made up altogether of affectionate feeling. What I saw of him in private gave me the most exalted opinion of him as a Christian. Oh, that such men as Mr. Wilberforce were more common in this world. So much human blood would not then be shed to gratify the malice and revenge of a few wicked, interested men.

"I hope Cousin Samuel Breese will distinguish himself under so gallant a commander as Captain Perry. I shall look with anxiety for the sailing of the Guerrière. There will be plenty of opportunity for him, for peace with us is deprecated by the people here, and it only remains for us to fight it out gallantly, as we are able to do, or submit slavishly to any terms which they please to offer us. A number of *humane* schemes are under contemplation, such as burning New London for the sake of the frigates there; arming the blacks in the Southern States; burning all of our principal cities, and such like plans, which, from the supineness of the New England people, may be easily carried into effect. But no, the *humane, generous* English cannot do such base things — I hope not; let the event show it. It is perhaps well I am here, for, with my present opinions, if I were at home, I should most certainly be in the army or navy. My mite is small, but, when my country's honor demands it, it might help to sustain it.

"There can now be no French party. I wish very much to know what effect this series of good news will have at home. I congratulate you as well as all other good people on the providential events which have lately happened; they must produce great changes with us; I hope it will be for the best.

"I am in excellent health, and am painting away; I am making studies for the large picture I contemplate for next year. It will be as large, I think, as Mr. Allston's famous one, which was ten feet by fourteen."

It can hardly be wondered at that the parents should have been somewhat anxious, when we learn from letters of June, 1814, that they had not heard from their son for *seven months.* They were greatly relieved when letters did finally arrive, and they rejoiced in his success and in the hope of a universal peace, which should enable their sons "to act their part on the stage of life in a calmer period of the world."

His mother keeps urging him to send some of his paintings home, as they wish to judge of his improvement, having, as yet, received nothing but the small pen-and-ink portrait of himself, which they do not think a very good likeness. She also emphatically discourages any idea of patronage from America, owing to the hard times brought on by the war, and the father tells his son that he will endeavor to send him one thousand dollars more, which must suffice for the additional year's study and the expenses of the journey home.

It is small wonder that the three sons always manifested the deepest veneration and affection for their parents, for seldom has there been seen as great devotion and self-sacrifice, and seldom were three sons more worthy of it. Sidney was at this time studying law at Litchfield, Connecticut, and Richard was attending the Theological Seminary at Andover, Massachusetts. Both became eminent in after life, though, curiously enough, neither in the law nor in the ministry. But we shall have occasion to treat more specifically of this later on. The three brothers were devotedly attached to each other to the very end of their long lives, and were mutually helpful as their lives now diverged and now came together again.

The next letter from Morse to his parents, written on June 15, 1814, gives a further account of the great people who were at that time in London: —

"I expected at this time to have been in Bristol with Mr. and Mrs. Allston, who are now there, but the great fêtes in honor of the peace, and the visit of the allied sovereigns, have kept me in London till all is over. There are now in London upward of twenty foreign princes; also the great Emperor Alexander and the King of Prussia. A week ago yesterday they arrived in town, and, contrary to expectation, came in a very private manner. I went to see their *entrée*, but was disappointed with the rest of the people, for the Emperor Alexander, disliking all show and parade, came in a private carriage and took an indirect route here.

"The next and following day I spent in endeavoring to get a sight of them. I have been very fortunate, having seen the Emperor Alexander no less than fourteen times, so that I am quite familiar with his face; the King of Prussia I have seen once; Marshal Blücher, five or six times; Count Platoff, three or four times; besides Generals de Yorck, Bülow, etc., all whose names must be perfectly familiar to you, and the distinguished parts they have all acted in the great scenes just past.

"The Emperor Alexander I am quite in love with; he has every mark of a great mind. His countenance is an uncommonly fine one; he has a fair complexion, hair rather light, and a stout, well-made figure; he has a very cheerful, benevolent expression, and his conduct has everywhere evinced that his face is the index of his mind. When I first saw him he was dressed in a green uniform with two epaulets and stars of different orders; he was conversing at the window of his hotel with his sister, the Duchess of Oldenburg. I saw him again soon after in the superb coach of the Prince Regent, with the Duchess, his sister, going to the court of the Queen. In a few hours after I saw him again on the balcony of the Pultney Hotel; he came forward and bowed to the people. He was then dressed in a red uniform, with a broad blue sash over the right shoulder; he appeared to great advantage; he stayed about five minutes. I saw him again five or six times through the day, but got only indifferent views of him. The following day, however, I was determined to get a better and nearer view of him than before. I went down to his hotel about ten o'clock, the time when I supposed he would leave it; I saw one of the Prince's carriages drawn up, which opened at the top and was thrown back before and behind. In a few minutes the Emperor with his sister made their appearance and got into it. As the carriage started, I pressed forward and got hold of the ring of the coach door and kept pace with it for about a quarter of a mile. I was so near that I could have touched him; he was in a plain dress, a brown coat, and altogether like any other gentleman. His sister, the Duchess, also was dressed in a very plain, unattractive manner, and, if it had not been for the crowd which followed, they would have been taken for any lady and gentleman taking an airing.

"In this unostentatious manner does he conduct himself, despising all pomp, and seems rather more intent upon inspecting the charitable, useful, and ornamental establishments of this country, with a view, probably, of benefiting his own dominions by his observations, than of displaying his rank by the splendor of dress and equipage.

"His condescension also is no less remarkable. An instance or two will exemplify it. On the morning after his arrival he was up at six o'clock, and, while the lazy inhabitants of this great city were fast asleep in their beds, he was walking with his sister, the Duchess, in Kensington Gardens. As he came across Hyde Park he observed a corporal drilling some recruits, upon which he went up to him and entered into familiar conversation with him, asking him a variety of questions, and, when he had seen the end of the exercise, shook him heartily by the hand and left him. When he was riding on horseback, he shook hands with all who came round him.

"A few days ago, as he was coming out of the gate of the London Docks on foot, after having inspected them, a great crowd was waiting to see him, among whom was an old woman of about seventy years of age, who seemed very anxious to get near him, but, the crowd pressing very much, she exclaimed, 'Oh, if I could but touch his clothes!' The Emperor overheard her, and, turning round, advanced to her, and, pulling off his glove, gave her his hand, and, at the same time dropping a guinea into hers, said to her, 'Perhaps this will do as well.' The old woman was quite overcome, and cried, 'God bless Your Majesty,' till he was out of sight.

"An old woman in her ninetieth year sent a pair of warm woolen stockings to the Emperor, and with them a letter stating that she had knit them with her own hands expressly for him, and, as she could not afford to send him silk, she thought that woolen would be much more acceptable, and would also be more useful in his climate. The Emperor was very much pleased, and determined on giving her his miniature set in gold and diamonds, but, upon learning that her situation in life was such that money would be more acceptable, he wrote her an answer, and, thanking her heartily for her present, enclosed her one hundred pounds.

"These anecdotes speak more than volumes in praise of the Emperor Alexander. He is truly a great man. He is a great conqueror, for he has subdued the greatest country in the world, and overthrown the most alarming despotism that ever threatened mankind. He is great also because he is good; his whole time seems spent in distributing good to all around him; and whereever he goes he makes every heart rejoice. He is very active and is all the time on the alert in viewing everything that is worth seeing. The Emperor is also extremely partial to the United States; everything American pleases him, and he seems uncommonly interested in the welfare of our country.

"I was introduced to-day to Mr. Harris, our *chargé d'affaires* to the court of Russia. He is a very intelligent, fine man, and is a great favorite with Alexander. From a conversation with him I have a scheme in view which, when I have matured, I will submit to you for your approbation.

"The King of Prussia I have seen but once, and then had but an imperfect view of him. He came to the window with the Prince Regent and bowed to the people (at St. James's Palace). He is tall and thin, has an agreeable countenance, but rather dejected in consequence of the late loss of his queen, to whom he was very much attached.

"General Blücher, now Prince Blücher, I have seen five or six times. I saw him on his entrance into London, all covered with dust, and in a very ordinary kind of vehicle. On the day after I saw him several times in his carriage, drawn about wherever he wished by the *mob*. He is John's greatest favorite, and they have almost pulled the brave general and his companion, Count Platoff, to pieces out of pure affection. Platoff had his coat actually torn off him and divided into a thousand pieces as *relics* by the good people — their kindness knows no bounds, and, I think, in all the battles which they have fought, they never have run so much risk of losing their limbs as in encountering their friends in England.

"Blücher is a veteran-looking soldier, a very fine head, monstrous mustaches. His head is bald, like papa's, his hair gray, and he wears powder. Understanding that he was to be at Covent Garden Theatre, I went, as the best place to see him, and I was not disappointed. He was in the Prince's box, and I had a good view of him during the whole entertainment, being directly before him for three or four hours. A few nights since I also went to the theatre to see Platoff, the *hetman* (chief) of the Cossacks. He has also a very fine countenance, a high and broad forehead, dark complexion, and dark hair. He is tall and well-made, as I think the Cossacks are generally. He was very much applauded by a crowded house, the most part collected to see him."

The following letter is from Washington Allston written in Bristol, on July 5, 1814: —

MY DEAR SIR, — I received your last on Saturday and should have answered your first letter but for two reasons.

First, that I had nothing to say; which, I think, metaphysicians allow to be the most natural as well as the most powerful cause of silence.

Second, that, if I had had anything to say, the daily expectation which I entertained of seeing you allowed no confidence in the hope that you would hear what I had to say should I have said it.

I thank you for your solicitude, and can assure you that both Mrs. Allston and myself are in every respect better than when we left London. Mr. King received me, as I wished, with undiminished kindness, and was greatly pleased with the pictures. He has not, however, seen the large one, which, to my agreeable surprise, I have been solicited from various quarters to exhibit, and that, too, without my having given the least intimation of such a design. I have taken Merchant Tailors' Hall (a very large room) for this purpose, and shall probably open it in the course of next week.

Perhaps you will be surprised to hear that I have been retouching it. I have just concluded a fortnight's hard work upon it, and have the satisfaction to add that I have been seldom better satisfied than with my present labor. I have repainted the greater part of the draperies — indeed, those of all the principal figures, excepting the Dead Man — with powerful and positive colors, and added double strength to the shadows of every figure, so that for force and distinctness you would hardly know it for the same picture. The "Morning Chronicle" would have no reason now to complain of its "wan red." . . .

I am sorry that Parliament has been so unpolite to you in procrastinating the fireworks. But they are an unpolished set and will still be in the dark age of incivility notwithstanding their late illuminations. However I am in great hopes that the good people of England will derive no small degree of moral embellishment from their pure admiration of the illustrious General B——, who, it is said, for drinking and gaming has no equal.

BRISTOL, September 9, 1814.

MY DEAR PARENTS, — Your kind letters of June last I have received, and return you a thousand thanks for them. They have relieved me from a painful state of anxiety with respect to my future prospects. I cannot feel too thankful for such kind parents who have universally shown so much indulgence to me. Accept my gratitude and love; they are all I can give.

You allow me to stay in Europe another year. Your letters are not in answer to some I have subsequently sent requesting leave to reside in Paris. Mr. Allston, as well as all my friends, think it by all means necessary I should lose no time in getting to France to improve myself for a year in drawing (a branch of art in which I am very deficient).

I shall therefore set out for Paris in about two weeks, unless your letters in answer to those sent by Drs. Heyward and Cushing should arrive and say otherwise. Since coming to Bristol I have not found my prospects so good as I before had reason to expect (owing in a great degree to political irritation). I have, however, contrived to make sufficient to pay off *all* my *debts*, which have given me some considerable uneasiness.

I can live much more reasonably in Paris (indeed, some say for half what I can in London); I can improve myself more; and, therefore, all things taken into consideration, I believe it would be agreeable to my parents. As to the political state of Paris, there is nothing to fear from that. It appears perfectly tranquil, and should at any time any difficulties arise, it is but three days' journey back to England again. Besides this, I hope my parents will not feel any solicitude for me lest I should fall into any bad way, when they consider that I am now between twenty-three and twenty-four years of age, and that this is an age when the habits are generally fixed.

As for expense, I must also request your confidence. Feeling as I do the great obligations I am under to my parents, they must think me destitute of gratitude if they thought me capable, after all that has been said to me, of being prodigal. The past I trust you will find to be an example for the future.

In a letter from a friend, M. Van Schaick, written from Dartmouth, October 13, 1814, after speaking in detail of the fortifications of New York Harbor, which he considers "impregnable," we find the following interesting information: —

"But what satisfies my mind more than anything else is that all the heights of Brooklyn on Long Island are occupied by strong chains of forts; the Captain calls it an iron-work; and that the steamboat frigate, carrying forty-four 32-pounders, must by this time be finished. Her sides are eight feet thick of solid timber. No ball can penetrate her. . . . The steamboat frigate is 160 feet long, 40 wide, carries her wheels in the centre like the ferry-boats, and will move six miles an hour against a common wind and tide. She is the wonder and admiration of all beholders."

From this same gentleman is the following letter, dated October 21, 1814: —

My dear Friend, — My heart is so full that I do not know how to utter its emotions. Thanks, all thanks to Heaven and our glorious heroes! My satisfaction is full; it is perfect. It partakes of the character of the victory and wants nothing to make it complete.

I return your felicitations upon this happy and heart-cheering occasion, and hope it may serve to suppress every sigh and to enliven every hope that animates the bosoms of my friends at Bristol. Give Mr. Allston a hearty squeeze of the hand for me in token of my gratification at this event and my remembrance of him.

I enter into your feelings; I enjoy your triumph as much as if I was with you. May it do you good and lengthen your lives. Really I think it is much more worth my regard to live now than ever it was before. This gives a tone to one's nerves, a zest to one's appetite, and a reality to existence that pervades all nature and exhibits its effects in every word and action.

Among the heroes whose names shall be inscribed upon the broad base of American Independence and Glory, the names of the heroes of Lake Erie and Lake Champlain will be recognized as brilliant and every way worthy; and it will hereafter be said that the example and exertions of New York have saved the nation. . . . What becomes of Massachusetts now and its sage politicians? Oh! shut the picture; I cannot bear the contrast. Like a dead carcass she hangs upon the living spirit which animates the heart, and she impedes its motions. Her consequence is gone, and I am sorry for it, because I have been accustomed to admire the noble spirit she once displayed, and the virtues which adorned her brighter days. . . .

We sail on Sunday or Monday. I have received the box. Everything is right. Heaven bless you.

Going back a few days in point of time, the following letter was written to his parents: —

BRISTOL, October 11, 1814.

Your letters to the 31st of August have been received, and I have again to express to you my thanks for the sacrifices you are making for me. One day I hope it will be in my power to repay you for the many acts of indulgence to me. . . .

Your last letters mention nothing about my going to France. I perceive you have got my letters requesting leave, but you are altogether silent on the subject. Everything is in favor of my going, my improvement, my expenses, and, last though not least, *the state of my feelings*. I shall be ruined in my feelings if I stay longer in England. I cannot endure the continued and daily insults to my feelings as an American. But on this head I promised not to write anything more; still allow me to say but a few words — On second thoughts, however, I will refer you entirely to Dr. Romeyn. If it is possible, as you value my comfort, see him as speedily as possible. He will give you my sentiments exactly, and I fully trust that, after you have heard him converse for a short time, you will completely liberate me from the imputation of error. . . .

Mr. Bromfield [the merchant through whom he received his allowance] thinks I had better wait until I receive positive leave from you to go to France. Do write me soon and do give me leave. I long to bury myself in the Louvre in a country at least not hostile to mine, and where guns are not firing and bells ringing for victory over my countrymen. . . . Where is American patriotism, — how long shall England, already too proud, glory in the blood of my countrymen? Oh! for the genius of Washington! Had I but his talents with what alacrity would I return to the relief of that country which (without affectation, my dear parents) is dearer to me than my life. Willingly (I speak with truth and deliberation), willingly would I sacrifice my life for her honor.

Do not think ill of me for speaking thus strongly. You cannot judge impartially of my feelings until you are placed in my situation. Do not say I suffer myself to be carried away by my feelings; your feelings could never have been tried as mine have; you cannot see with the eyes I do; you cannot have the means of ascertaining facts on this side of the water that I have. But I will leave this subject and only say *see Dr. Romeyn*. . . .

I find no encouragement whatever in Bristol in the way of my art. National feeling is mingled with everything here; it is sufficient that I am an American, a title I would not change with the greatest king in Europe.

I find it more reasonable, living in Bristol, or I should go to London immediately. Mr. and Mrs. Allston are well and send you their respects. They set out for London in a few days after some months' *unsuccessful* (between ourselves) residence here. All public feeling is absorbed in one object, the *conquest of the United States;* no time to encourage an artist, especially an American artist.

I am well, extremely well, but not in good spirits, as you may imagine from this letter. I am painting a little landscape and am studying in my mind a great historical picture, to be painted, by your leave, in Paris.

CHAPTER VIII

NOVEMBER 9, 1814 — APRIL 23, 1815

Does not go to Paris. — Letter of admonition from his mother. — His parents' early economies. — Letter from Leslie. — Letter from Rev. S. F. Jarvis on politics. — The mother tells of the economies of another young American, Dr. Parkman. — The son resents constant exhortations to economize, and tells of meanness of Dr. Parkman. — Writes of his own economies and industry. — Disgusted with Bristol. — Prophesies peace between England and America. — Estimates of Morse's character by Dr. Romeyn and Mr. Van Schaick. — The father regrets reproof of son for political views. — Death of Mrs. Allston. — Disagreeable experience in Bristol. — More economies. — Napoleon I. — Peace.

MORSE did not go to Paris at this time. The permission from his parents was so long delayed, owing to their not having received certain letters of his, and his mentor, Mr. Bromfield, advising against it, he gave up the plan, with what philosophy he could bring to bear on the situation.

His mother continued to give him careful advice, covering many pages, in every letter. On November 9, 1814, she says: —

"We wish to know what the plan was that you said you were maturing in regard to the Emperor of Russia. You must not be a schemer, but determine on a steady, uniform course. It is an old adage that 'a rolling stone never gathers any moss'; so a person that is driving about from pillar to post very seldom lays up anything against a rainy day. You must be wise, my son, and endeavor to get into such steady business as will, with the divine blessing, give you a support. Secure that first, and then you will be authorized to indulge your taste and exercise your genius in other ways that may not be immediately connected with a living.

"You mention patronage from this country, but such a thing is not known here unless you were on the spot, and not then, indeed, but for value received. You must therefore make up your mind to labor for yourself without leaning on any one, and look up to God for his blessing upon your endeavors. This is the way your parents set out in life about twenty-five years ago. They had nothing to look to for a support but their salary, which was a house, twenty cords of wood, and $570 a year. The reception and circulation of the Geography was an experiment not then made. With the blessing of Heaven on these resources we have maintained an expensive family, kept open doors for almost all who chose to come and partake of our hospitality. Enemies, as well as friends, have been welcomed. We have given you and your brothers a liberal education, have allowed you $4000, are allowing your brothers about $300 a year apiece, and are supporting our remaining family at the rate of $2000 a year. This is a pretty correct statement, and I make it to show you what can be done by industry and economy, with the blessing of Heaven."

While Morse was in Bristol, his friend C. R. Leslie thus writes to him in lead pencil from London, on November 29, 1814: —

MOST POTENT, GRAVE AND REVEREND DOCTOR, — I take up my pencil to make ten thousand apologies for addressing you in humble black lead. Deeply impressed as I am with the full conviction that you deserve the very best Japan ink, the only excuse I can make to you is the following. It is, perhaps, needless to remind you that the tools with which ink is applied to paper, in order to produce writing, are made from goose quills, which quills I am goose enough not to keep a supply of; and not having so much money at present in my breeches pocket as will purchase one, I am forced to betake myself to my pencil; an instrument which, without paying myself any compliment, I am sure I can wield better than a pen.

I am glad to hear that you are so industrious, and that Mr. Allston is succeeding so well with portraits. I hope he will bring all he has painted to London. I am looking out for you every day. I think we form a kind of family here, and I feel in an absence from Mr. and Mrs. Allston and yourself as I used to do when away from my mother and sisters.

By the bye, I have not had any letters from home for more than a month. It seems the Americans are all united and we shall now have war in earnest. I am glad of it for many reasons; I think it will not only get us a more speedy and permanent peace, but may tend to crush the demon of party spirit and strengthen our government.

I am done painting the gallery, and have finished my drawings for the frieze. Thank you for your good wishes.

I thought Mr. Allston knew how proud I am of being considered his student. Tell him, if he thinks it worth while to mention me at all in his letter to Delaplaine, I shall consider it a great honor to be called his student.

The father, in a letter of December 6, 1814, after again urging him to leave politics alone, adds this postscript:

"P.S. If you can make up your mind to remain in

London and finish your great picture for the exhibition; to suppress your political feelings, and resolutely turn a deaf ear to everything which does not concern your professional studies; not to talk on politics and preserve a conciliating course of conduct and conversation; make as many friends as you can, and behave as a good man ought to in your situation, and put off going to France till after your exhibition, — this plan would suit us best. But with the observations and advice now before you, we leave you to judge for yourself. Let us early know your determination and intended plans. You must rely on your own resources after this year."

The following letter is from his warm friend, the Reverend Samuel F. Jarvis, written in New York, December 14, 1814: —

"I am not surprised at the feelings you express with regard to England or America. The English in general have so contemptuous an opinion of us and one so exalted of themselves, that every American must feel a virtuous indignation when he hears his country traduced and belied. But, my dear sir, it is natural, on the other hand, for an exile from his native land to turn with fond remembrance to its excellences and forget its defects. You will be able some years hence to speak with more impartiality on this subject than you do at present.

"The men who have involved the country in this war are wicked and corrupt. A systematic exclusion of all Federalists from any office of trust is the leading feature of this Administration, yet the Federalists comprehend the majority of the wealth, virtue, and intelligence of the community. It is the power of the ignorant multitude by which they are supported, and I conceive that

America will never be a respectable nation in the eyes of the world, till the extreme democracy of our Constitution is done away with, and there is a representation of the property rather than of the population of the country. You feel nothing of the oppressive, despotic sway of the *soi-disant* Republicans, but we feel it in all its bitterness, and know that it is far worse than that of the most despotic sovereigns in Europe. With such men there can be no union.

"The repulsion of British invasion is the duty, and will be the pride, of every American; but, while prepared to bare his arm in defence of his much-wronged country against a proud and arrogant, and, in some instances, a cruel, foe, he cannot be blind to the unprincipled conduct of her internal enemies, and such he must conceive the present ruling party to be."

On December 19, 1814, his mother writes: —

"I was not a little astonished to hear you say, in one of your letters from Bristol, that you had earned money enough there to pay off your debts. I cannot help asking what debts you could have to discharge with your own earnings after receiving one thousand dollars a year from us, which we are very sure must have afforded you, even by your own account of your expenses, ample means for the payment of all just, fair, and honorable debts, and I hope you contract no others. We are informed by others that they made six hundred dollars a year not only pay all their expenses of clothing, board, travelling, learning the French language, etc., etc., but they were able out of it to purchase books to send home, and actually sent a large trunk full of elegant books. Now the person who told us that he did this has a father who is said to be worth a hundred and fifty thousand dollars; therefore the young man was not pinched for means, but was thus economical out of consideration to his parents, and to show his gratitude to them, as I suppose. Now think, my dear son, how much more your poor parents are doing for you, how good your dear brothers are to be satisfied with so little done for them in comparison with what we are doing for you, and let the thought stimulate you to more economy and industry. I greatly fear you have been falling off in both these since the éclat you received for your first performances. It has always been a failing of yours, as soon as you found you could excel in what you undertook, to be tired of it and not trouble yourself any further about it. I was in hopes that you had got over this fickleness ere this. . .

"You must not expect to paint anything in this country, for which you will receive any money to support you, but portraits; therefore do everything in your power to qualify you for painting and taking them in the best style. That is all your hope here, and to be very obliging and condescending to those who are disposed to employ you. . . .

"I think young Leslie is a very estimable young man to be, as I am told he is, supporting himself and assisting his widowed mother by his industry."

I shall anticipate a little in order to give at once the son's answer to this reproof. He writes on April 23, 1815:—

"I wish I could persuade my parents that they might place some little confidence in my judgment at the age I now am (nearly twenty-four), an age when, in ordinary people, the judgment has reached a certain degree of maturity. It is a singular and, I think, an unfortunate fact that I have not, that I recollect, since I have been in England, had a turn of low spirits except when I have received letters from home. It is true I find a great deal of affectionate solicitude in them, but with it I also find so much complaint and distrust, so much fear that I am doing wrong, so much doubt as to my morals and principles, and fear lest I should be led away by bad company and the like, that, after I have read them, I am miserable for a week. I feel as though I had been guilty of every crime, and I have passed many sleepless nights after receiving letters from you. I shall not sleep to-night in consequence of passages in your letters just received."

Here he quotes from his mother's letter and answers:

"Now as to the young man's living for six hundred dollars, I know who it is of whom you speak. It is Dr. Parkman, who made it his boast that he would live for that sum, but you did not enquire *how* he lived. I can tell you. He never refused an invitation to dine, breakfast, or tea, which he used to obtain often by pushing himself into everybody's company. When he did not succeed in getting invitations, he invited himself to breakfast, dine, or sup with some of his friends. He has often walked up to breakfast with us, a distance of three or four miles. If he failed in getting a dinner or meal at any of these places, he either used to go without, or a bit of bread answered the purpose till next meal. In his dress he was so shabby and uncouth that any decent person would be ashamed to walk with him in the street. Above all, his notorious meanness in his money matters, his stickling with his poor washerwoman for a halfpenny and with others for a farthing, and his uniform stinginess on all occasions rendered him notoriously disgusting to all his acquaintances, and affords, I should imagine, but a poor example for imitation. . . .

"The fact is I could live for *fifty* pounds a year if my only object was to live cheap, and, on the other hand, if I was allowed one thousand pounds a year, I could spend it all without the least extravagance in obtaining greater advantages in my art. But as your goodness has allowed me but two hundred pounds (and I wish you again to receive my sincere thanks for this allowance), should not my sole endeavor be to spend all this to the utmost advantage; to keep as closely within the bounds of that allowance as possible, and would not *economy* in this instance consist in rigidly keeping up to this rule? If this is a true statement of the case, then have I been perfectly economical, for I have not yet overrun my allowance, and I think I shall be able to return home without having exceeded it a single shilling. If I have done this, and still continue to do it, why, in every letter I receive from home, is the injunction repeated of *being economical?* It makes me exceedingly unhappy, especially when I am conscious of having used my utmost endeavors, ever since I have been in England, to be rigidly so.

"As to *industry,* in which mama fears I am falling off, I gave you an account in my last letter (by Mr. Ralston) of the method I use in parcelling out my time. Since writing that letter the spring and summer are approaching fast, and the days increasing. Of course I can employ more of the time than in the winter. Mr. Leslie and myself rise at five o'clock in the morning and walk about a mile and a half to Burlington, where are the famous

Elgin Marbles, the works of Phidias and Praxiteles, brought by Lord Elgin from Athens. From these we draw three hours every morning, wet or dry, before breakfast, and return home just as the bustle begins in London, for they are late risers in London. When we go out of a morning we meet no one but the watchman, who goes his rounds for an hour and a half after we are up. Last summer Mr. Leslie and I used to paint in the open air in the fields three hours before breakfast, and often before sunrise, to study the morning effect on the landscape.

"Now, being conscious of employing my time in the most industrious manner possible, you can but faintly conceive the mortification and sorrow with which I read that part of mama's letter. I was so much hurt that I read it to Mr. Allston, and requested he would write to you and give you an account of my spending my time. He seemed very much astonished when I read it to him, and *authorized me to tell you from him that it was impossible for any one to be more indefatigable in his studies than I am.*

"Mama mentions in her letter that she hears that Mr. Leslie supports his mother and sisters by his labors. This is not the case. Leslie was supported by three or four individuals in Philadelphia till within a few months past. About a year ago he sold a large picture which he painted (whilst I was on my fruitless trip to Bristol for money) for a hundred guineas. Since that he has had a number of commissions in portraits and is barely able to support himself; indeed, he tells me this evening that he has but £20 left. He is a very economical and a most excellent young man. His expenses in a year are, on an average, from £230 to £250; Mr. Allston's (single) expenses not less than £300 per annum, and I know of no artist among all my acquaintance whose expenses in a year are less than £200."

Returning now to the former chronological order, I shall include the following vehement letter written from London on December 22, 1814: —

My dear Parents, — I arrived yesterday from Bristol, where I have been for several months past endeavoring to make a little in the way of my profession, but have completely failed, owing to several causes.

First, the total want of anything like partiality for the fine arts in that place; the people there are but a remove from brutes. A "Bristol hog" is as proverbial in this country as a "Charlestown gentleman" is in Boston. Their whole minds are absorbed in trade; barter and gain and interest are all they understand. If I could have painted a picture for half a guinea by which they could have made twenty whilst I starved, *I could have starved.*

Secondly, the virulence of national prejudice which rages now with tenfold acrimony. They no longer despise, they hate, the Americans. The battle on Champlain and before Plattsburgh has decided the business; the moans and bewailings for this business are really, to an American, quite comforting after their arrogant boasting of reducing us to unconditional submission.

Is it strange that I should feel a little the effects of this universal hatred? I have felt it, and I have left Bristol after six months' perfect neglect. After having been invited there with promises of success, I have had the mortification to leave it without having, from Bristol, a single commission. More than that, and by far the worst, if I have not gone back in my art these six months, I have at least stood still, and to me this is the most trying reflection of all. I have been immured in the paralyzing atmosphere of trade till my mind was near partaking the infection. I have been listening to the grovelling, avaricious devotees of mammon, whose souls are narrowed to the studious contemplation of a hard-earned shilling, whose leaden imaginations never soared above the prospect of a good bargain, and whose *summum bonum* is the inspiring idea of counting a hundred thousand: I say I have been listening to these miserly beings till the idea did not seem so repugnant of lowering my noble art to a trade, of painting for money, of degrading myself and the soul-enlarging art which I possess, to the narrow idea of merely getting money.

Fie on myself! I am ashamed of myself; no, never will I degrade myself by making a trade of a profession. If I cannot live a gentleman, I will starve a gentleman. But I will dismiss this unpleasant subject, the particulars of which I can better relate to you than write. Suffice it to say that my ill-treatment does not prey upon my spirits; I am in excellent health and spirits and have great reason to be thankful to Heaven for thousands of blessings which one or two reverses shall not make me forget. Reverses do I call them? How trifling are my troubles to the millions of my fellow creatures who are afflicted with all the dreadful calamities incident to this life. Reverses do I call them? No, they are blessings compared with the miseries of thousands.

Indeed, I am too ungrateful. If a thing does not result

just as I wish, I begin to repine; I forget the load of blessings which I enjoy: life, health, parents whose kindness exceeds the kindest; brothers, relatives, and friends; advantages which no one else enjoys for the pursuit of a favorite art, besides numerous others; all which are forgotten the moment an unpleasant disappointment occurs. I am very ungrateful.

With respect to peace, I can only say I should not be surprised if the preliminaries were signed before January. My reasons are that Great Britain cannot carry on the war any longer. She may talk of her inexhaustible resources, but she well knows that the great resource, the property tax, must fail next April. The people will not submit any longer; they are taking strong measures to prevent its continuance, and without it they cannot continue the war.

Another great reason why I think there will be peace is the absolute *fear* which they express of us. They fear the increase of our navy; they fear the increase of the army; they fear for Canada, and they are in dread of the further disgrace of their national character. Mr. Monroe's plan for raising 100,000 men went like a shock through the country. They saw the United States assume an attitude which they did not expect, and the same men who cried for "war, war," "thrash the Americans," now cry most lustily for peace.

The union of the parties also has convinced them that we are determined to resist their most arrogant pretensions.

Love to all, brothers, Miss Russell, etc.

Yours very affectionately,

SAML. F. B. MORSE.

He ends the letter thus abruptly, probably realizing that he was beginning to tread on forbidden ground, but being unable to resist the temptation.

While from this letter and others we can form a just estimate of the character and temperament of the man, it is also well to learn the opinion of his contemporaries; I shall, therefore, quote from a letter to the elder Morse of the Dr. Romeyn, whom the son was so anxious to have his father see, also from a letter of Mr. Van Schaick to Dr. Romeyn.

The former was written in New York, on December 27, 1814.

"The enclosed letter of my friend Mr. Van Schaick will give you the information concerning your son which you desire. He has been intimately acquainted with your son for a considerable time. You may rely on his account, as he is not only a gentleman of unquestionable integrity, but also a professor of the Lord Christ. What I saw and heard of your son pleased me, and I cannot but hope he will repay all your anxieties and realize your reasonable expectations by his conduct and the standing which he must and will acquire in society by that conduct."

Mr. Van Schaick's letter was written also in New York, on December 14, 1814: —

"To those passages of Dr. Morse's letter respecting his son, to which you have directed my attention, I hasten to reply without any form, because it will gratify me to relieve the anxiety of the parents of my friend. His religious and moral character is unexceptionally good. He feels strongly for his country and expresses those feelings among his American friends with great sensibility. I do not know that he ever indulges in any observations in the company of Englishmen which are calculated to injure his standing among them. But, my dear sir, you fully know that an American cannot escape the sting of illiberal and false charges against his country and even its moral character, unless he almost entirely withholds himself from society. It cannot be expected that any human being should be so unfeeling as to suffer indignity in total silence.

"But I do not think that any political collisions, which may incidentally and very infrequently arise, can injure him as an artist; for it is well known to you that the simple fact of his being an American is sufficient to prevent his rising rapidly into notice, since the possession of that character clogs the efforts, or, at least, somewhat clouds the fame of men of superior genius and established talent. . . . I advised Samuel to go to France and bury himself for six months in the Louvre; from thence to Italy, the seat of the arts. He inclined to the first part of the plan, and then to return home, but deferred putting it into execution till he heard from his father. Mr. Allston intended to winter in London. Morse has a fine taste and colors well. His drawing is capable of much improvement, but he is anxious to place himself at the head of his profession, and, with a little judicious encouragement, will probably succeed. That patient industry which has in all ages characterized the masters of the art, he will find it to his interest to apply to his studies the farther he advances in them. His success has been moderately good. If he could sell the pictures he has on hand, the avails would probably pay his way into France."

Referring to these letters the father, writing on January 25, 1815, says: —

"We have had letters from Dr. Romeyn and Mr. Van Schaick concerning you which have comforted us much. Since receiving them we don't know but we have expressed ourselves, in our letters in answer to your last, a little stronger than we ought in regard to your *political* feelings and conduct. I find others who have returned feel pretty much as you do. But it should be remembered that your situation as an artist is different from theirs. It is your wisdom to leave politics to politicians and be solely the artist. But if you are in France these cautions will probably not be necessary, as you will have no temptation to enter into any political discussions."

On the 3d of February, 1815, Morse, in writing to his parents, has a very sad piece of news to communicate to them: —

"I write in great haste and much agitation. Mrs. Allston, the wife of our beloved friend, died last evening, and the event overwhelmed us all in the utmost sorrow. As for Mr. Allston, for several hours after the death of his wife he was almost bereft of reason. Mr. Leslie and I are applying our whole attention to him, and we have so far succeeded as to see him more composed."

This was a terrible grief to all the little coterie of friends, for whom the Allston house had been a home. One of them, Mr. J. J. Morgan, in a long letter to Morse written from Wiltshire, thus expresses himself: —

"Gracious God! unsearchable, indeed, are thy ways! The insensible, the brutish, the wicked are powerful and everywhere, in everything successful; while Allston, who is everything that is amiable, kind, and good, has been bruised, blow after blow, and now, indeed, his cup is full. I am too unwell, too little recovered from the effect of your letter, to write much. Coleridge intends writing to-day; I hope he will. Allston may derive some little relief from knowing how much his friends partake of his grief."

This was a time of great discouragement to the young artist. Through the failure of some of his letters to reach his parents in time, he had not received their permission to go to France until it was too late for him to go. The death of Mrs. Allston cast a gloom over all the little circle, and, to cap the climax, he was receiving no encouragement in his profession. On March 10, 1815, he writes: —

"My jaunt to Bristol in quest of money completely failed. When I was first there I expected, from the little connection I got into, I should be able to support myself. I was obliged to come to town on account of the exhibitions, and stayed longer than I expected, intending to return to Bristol. During this time I received two pressing letters from Mr. Visscher (which I will show you), inviting me to come down, saying that I should have plenty of business. I accordingly hurried off. A gentleman, for whom I had before painted two portraits, had promised, if I would let him have them for ten guineas apiece, twelve being my price, that he would procure me five sitters. This I acceded to. I received twenty guineas and have heard nothing from the man since, though I particularly requested Mr. Visscher to enquire and remind him of his promise. Yet he never did anything more on the subject. I was there three months, gaining nothing in my art and without a single commission. Mr. Breed, of Liverpool, then came to Bristol. He took two landscapes which I had been amusing myself with (for I can say nothing more of them) at ten guineas each. I painted two more landscapes which are unsold.

"Mr. Visscher, a man worth about a hundred thousand pounds, and whose annual expenses, with a large family of seven children, are not one thousand, had a little frame for which he repeatedly desired me to paint a picture. I told him I would as soon as I had finished one of my landscapes. I began it immediately, without his knowing it, and determined to surprise him with it. I also had two frames which fitted Mr. Breed's pictures, and which I was going to give to Mr. Breed with his pictures. But Mr. Visscher was particularly pleased with the frames, as they were a pair, and told me not to send them to Mr. Breed as he should like to have them himself, and wished I would paint him pictures to fit them (the two other landscapes before mentioned). I accordingly was employed three months longer in painting these three pictures. I finished them; he was very much pleased with them; all his family were very much pleased with them; all who saw them were pleased with them. But he *declined taking them* without even asking my price, and said that he had more pictures than he knew what to do with.

"Mr. and Mrs. Allston heard him say twenty times he wished I would paint him a picture for the frame. Mr. Allston, who knew what I was about, told him, no doubt, I would do it for him, and in a week after I had completed it. I had told Mr. Visscher also that I was considerably in debt, and that, when he had paid me for these pictures, I should be something in pocket; and, by his not objecting to what I said, I took it for granted (and from his requesting me to paint the picture) that the thing was certain. But thus it was, without giving any reason in the world, except that he had pictures enough, he declined taking them, making me spend three months longer in Bristol than I otherwise should have done; standing still in my art, if not actually going back; and forcing me to run in debt for some necessary expenses of clothing in Bristol, and my passage from and back to London. During all this time not a single commission for a portrait, *many* of which were promised me, nor a single call from any one to look at my pictures. Thus ended my jaunt in quest of money.

"Do not think that this disappointment is in consequence of any misconduct of mine. Mr. Allston, who was with me, experienced the same treatment, and had it not been for his uncle, the American Consul, he might have starved for the Bristol people. His uncle was the only one who purchased any of his pictures. Since I have been in London I have been endeavoring to regain what I lost in Bristol, and I hope I have so far succeeded as to say: '*I have not gone back in my art.*'

"In order to retrench my expenses I have taken a painting-room out of the house, at about half of the expense of my former room. Though inconvenient in many respects, yet my circumstances require it and I willingly put up with it. As for *economy*, do not be at any more pains in introducing that personage to me. We have long been friends and necessary companions. If you could look in on me and see me through a day I think you would not tell me in every letter *to economize more*. It is impossible; I cannot economize more. I live on as plain food and as little as is for my health; less and plainer would make me ill, for I have given it a fair experiment. As for clothes, I have been decent and that is all. If I visited a great deal this would be a heavy expense, but, the less I go out, the less need I care for clothes, except for cleanliness. My only heavy expenses are colors, canvas, frames, etc., and these are heavy."

A number of pages of this letter are missing, much to my regret. He must have been telling of some of the great events which were happening on the Continent, probably of the Return from Elba, for it begins again abruptly.

"— when he might have avoided it by quietness; by undertaking so bold an attempt as he has done without being completely sure of success, and having laid his plans deeply; and, thirdly, I knew the feelings of the French people were decidedly in his favor, more especially the military. They feel as though Louis XVIII was forced upon them by their conquerors; they feel themselves a conquered nation, and they look to Bonaparte as the only man who can retrieve their character for them.

"All these reasons rushing into my mind at the time, I gave it as my opinion that Napoleon would again be Emperor of the French, and again set the world by the ears, unless he may have learned a lesson from his adversity. But this cannot be expected. I fear we are apt yet to see a darker and more dreadful storm than any we have yet seen. This is, indeed, an age of wonders.

"Let what will happen in Europe, let us have peace at home, among ourselves more particularly. But the character we have acquired among the nations of Europe in our late contest with England, has placed us on such high ground that none of them, England least of all, will wish to embroil themselves with us."

This was written just after peace had been established between England and America, and in a letter from his mother, written about the same time in March, 1815, she thus comments on the joyful news: "We have now the heartfelt pleasure of congratulating you on the return of peace between our country and Great Britain. May it never again be interrupted, but may both countries study the things that make for peace, and love as brethren."

It never has been interrupted up to the present day, for, as I am pursuing my pleasant task of bringing these letters together for publication, in the year of our Lord 1911, the newspapers are agitating the question of a fitting commemoration of a hundred years of peace between Great Britain and the United States.

Further on in this same letter the mother makes this request of her son: "When you return we wish you to bring some excellent black or corbeau cloth to make your good father and brothers each a suit of clothes. Your papa also wishes you to get made a handsome black cloth cloak for him; one that will fit you he thinks will fit him. Be sure and attend to this. Your mama would like some grave colored silk for a gown, if it can be had but for little. Don't forget that your mother is no dwarf, and that a large pattern suits her better than a small one."

The letter of April 23, from which I have already quoted, has this sentence at the beginning: "Your letters suppose me in Paris, *but I am not there;* you hope that I went in October last; I intended going and wished it at that time exceedingly, but I had not leave from you to go and Mr. Bromfield advised me by no means to go until I heard from you. You must perceive from this case how impossible it is for me to form plans, and transmit them across the Atlantic for approbation, thus letting an opportunity slip which is irrecoverable."

CHAPTER IX

MAY 3, 1815—OCTOBER 18, 1815

Decides to return home in the fall. — Hopes to return to Europe in a year. — Ambitions. — Paints "Judgment of Jupiter." — Not allowed to compete for premium. — Mr. Russell's portrait. — Reproof of his parents. — Battle of Waterloo. — Wilberforce. — Painting of "Dying Hercules" received by parents. — Much admired. — Sails for home. — Dreadful voyage lasting fifty-eight days. — Extracts from his journal. — Home at last.

It was with great reluctance that Morse made his preparations to return home. He thought that, could he but remain a year or two longer in an atmosphere much more congenial to an artist than that which prevailed in America at that time, he would surely attain to greater eminence in his profession.

He, in common with many others, imagined that, with the return of peace, an era of great prosperity would at once set in. But in this he was mistaken, for history records that just the opposite occurred. The war had made demands on manufacturers, farmers, and provision dealers which were met by an increase in inventions and in production, and this meant wealth and prosperity to many. When the war ceased, this demand suddenly fell off; the soldiers returning to their country swelled the army of the unemployed, and there resulted increased misery among the lower classes, and a check to the prosperity of the middle and upper classes. It would seem, therefore, that Fate dealt more kindly with the young man than he, at that time, realized; for, had he remained, his discouragements would undoubtedly have increased; whereas, by his return to his native land, although meeting with many disappointments and suffering many hardships, he was gradually turned into a path which ultimately led to fame and fortune.

On May 3, 1815, he writes to his parents: —

"With respect to returning home, I shall make my arrangements to be with you (should my life be spared) by the end of September next, or the beginning of October; but it will be necessary that I should be in England again (provided always Providence permits) by September following, as arrangements which I have made will require my presence. This I will fully explain when I meet you.

"The moment I get home I wish to begin work, so that I should like to have some portraits bespoken in season. I shall charge forty dollars less than Stuart for my portraits, so that, if any of my good friends are ready, I will begin the moment I have said 'how do ye do' to them.

"I wish to do as much as possible in the year I am with you. If I could get a commission or two for some large pictures for a church or public hall, to the amount of two or three thousand dollars, I should feel much gratified. I do not despair of such an event, for, through your influence with the clergy and their influence with their people, I think some commission for a scripture subject for a church might be obtained; a crucifixion, for instance.

"It may, perhaps, be said that the country is not rich enough to purchase large pictures; yes, but two or three thousand dollars can be paid for an entertainment which is gone in a day, and whose effects are to demoralize and debilitate, whilst the same sum expended on a fine picture would be adding an ornament to the country which would be lasting. It would tend to elevate and refine the public feeling by turning their thoughts from sensuality and luxury to intellectual pleasures, and it would encourage and support a class of citizens who have always been reckoned among the brightest stars in the constellation of American worthies, and who are, to this day, compelled to exile themselves from their country and all that is dear to them, in order to obtain a bare subsistence.

"I do not speak of *portrait-painters;* had I no higher thoughts than being a first-rate portrait-painter, I would have chosen a far different profession. My ambition is to be among those who shall revive the splendor of the fifteenth century; to rival the genius of a Raphael, a Michael Angelo, or a Titian; my ambition is to be enlisted in the constellation of genius now rising in this country; I wish to shine, not by a light borrowed from them, but to strive to shine the brightest.

"If I could return home and stay a year visiting my friends in various parts of the Union, and, by painting portraits, make sufficient to bring me to England again at the end of the year, whilst I obtained commissions enough to employ me and support me while in England, I think, in the course of a year or two, I shall have obtained sufficient credit to enable me to return home, if not for the remainder of my life, at least to pay a good long visit.

"In all these plans I wish you to understand me as always taking into consideration *the will of Providence;* and, in every plan for future operation, I hope I am not forgetful of the uncertainty of human life, and I wish always to say *should I live* I will do this or that. . . .

"I perceive by your late letters that you suppose I am painting a large picture. I did think of it some time ago and was only deterred on account of the expenses attending it. All this I will explain to your entire satisfaction when I see you, and why I do not think it expedient to make an exhibition when I return.

"I perceive also that you are a little too sanguine with respect to me and expect a little too much from me. You must recollect I am yet but a student and that a picture of any merit is not painted in a day. Experienced as Mr. West is (and he also paints quicker than any other artist), his last large picture cost him between three and four years' constant attention. Mr. Allston was nearly two years in painting his large picture. Young Haydon was three years painting his large picture, is now painting another on which he has been at work one year and expects to be two years more on it. Leslie was ten months painting his picture, and my 'Hercules' cost me nearly a year's study. So you see that large pictures are not the work of a moment.

"All these matters we will talk over one of these days, and all will be set right. I had better paint Miss Russell's, Aunt Salisbury's, and Dr. Bartlett's pictures at home for a very good reason I will give you."

He did, however, complete a large historical, or rather mythological, painting before leaving England. Whether it was begun before or after writing the foregoing letter, I do not know, but Mr. Dunlap (whom I have already quoted) has this to say about it: —

"Encouraged by the flattering reception of his first works in painting and in sculpture, the young artist redoubled his energies in his studies and determined to contend for the highest premium in historical composition offered by the Royal Academy at the beginning of the year 1814. The subject was 'The Judgment of Jupiter in the case of Apollo, Marpessa and Idas.' The premium offered was a gold medal and fifty guineas. The decision was to take place in December of 1815. The composition containing four figures required much study, but, by the exercise of great diligence, the picture was completed by the middle of July.

"Our young painter had now been in England four years, one year longer than the time allowed him by his parents, and he had to return immediately home; but he had finished his picture under the conviction, strengthened by the opinion of West, that it would be allowed to remain and compete with those of the other candidates. To his regret the petition to the council of the Royal Academy for this favor, handed in to them by West and advocated strongly by him and Fuseli, was not granted. He was told that it was necessary, according to the rules of the Academy, that the artist should be present to receive the premium; it could not be received by proxy. Fuseli expressed himself in very indignant terms at the narrowness of this decision.

"Thus disappointed, the artist had but one mode of consolation. He invited West to see his picture before he packed it up, at the same time requesting Mr. West to inform him through Mr. Leslie, after the premium should be adjudged in December, what chance he would have had if he had remained. Mr. West, after sitting before the picture for a long time, promised to comply with the request, but added: 'You had better remain, sir.'"

In a letter quoted, without a date, by Mr. Prime, which was written from Bristol, but which seems to have been lost, I find the following: —

"James Russell, Esq., has been extremely attentive to me. He has a very fine family consisting of four daughters and, I think, a son who is absent in the East Indies. The daughters are very beautiful, accomplished, and amiable, especially the youngest, Lucy. I came very near being at my old game of falling in love, but I find that love and painting are quarrelsome companions, and that the house of my heart is too small for both of them; so I have turned Mrs. Love out-of-doors. Time enough, thought I (with true old bachelor complacency), time enough for you these ten years to come. Mr. Russell's portrait I have painted as a present to Miss Russell, and will send it to her as soon as I can get an opportunity. It is an excellent likeness of him."

He must either have said more in this letter, or have written another after the family verdict (that terrible family verdict) had been pronounced, for in the letter of April 23, 1815, from which I have already quoted, he refers to this portrait as follows: —

"As to the portrait which I painted of Mr. Russell, I am sorry you mentioned it to Miss Russell, as I particularly requested that you would not, because, in case of failure, it would be a disappointment to her; but as you have told her, I must now explain. In the first place it is not a picture that will do me any credit. I was unfortunate in the light which I chose to paint him in; I wished to make it my best picture and so made it my worst, for I worked too timidly on it. It is a likeness, indeed, a very strong likeness, but the family are not pleased with it, and they say that I have not flattered him, that I have made him too old. So I determined I would not send it, indeed, I promised them I would not send it; but, notwithstanding, as I know Miss Russell will be good enough to comply with my conditions, I will send it directly; for, as it is a good likeness, every one except the family knowing it instantly, and Mr. Allston saying that it is *a very strong likeness*, it will on that account be a gratification to her. But I *particularly* and *expressly request* that it be kept in a private room to be shown *only* to friends and relations, and that *I may never be mentioned as the painter;* and, moreover, that no *artist* or *miniature painter* be allowed to see it. On these conditions I send it, taking for granted they will be complied with, and without waiting for an answer."

The parents of that generation were not frugal of counsel and advice, even when their children had reached years of discretion and had flown far away from the family nest.

The father, in a letter of May 20, 1815, thus gently reproves his son: —

"To-day we have received your letters to March 23. . . . You evidently misconceived our views in the letters to which you allude, and felt much too strongly our advice and remarks in respect to your writing us so much on politics. What we said was the affectionate advice of your parents, who loved you very tenderly, and who were not unwilling you should judge for yourself though you might differ from them. We have ever made a very candid allowance for you, and so have all your friends, and we have never for a moment believed we should differ a fortnight after you should come home and converse with us. You have, in the ardor of feeling, construed many observations in our letters as censuring you and designed to wound your feelings, which were not intended in the remotest degree by us for any such purpose. . . .

"I am sorry to hear of the death of Mr. Thornton. He was a good man."

His mother was much less gentle in her reproof. I cull the following sentences from a long letter of June 1, 1815: —

"In perfect consistency with the feelings towards you all, above described, we may and ought to tell you, and that with the greatest plainness, of anything that we deem improper in any part of your conduct, either in a civil, social, or religious view. This we feel it our duty to do and shall continue to do as long as we live; and it will ever be your duty to receive from us the advice, counsel, and reproof, which we may, from time to time, favor you with, with the most perfect respect and dutiful observance; and, when you differ from us on any point whatever, let that difference be conveyed to us in the most delicate and gentlemanly manner. Let this be done not only while you are under age and dependent on your parents for your support, but when you are independent, and when you are head of a family, and even of a profession, if you ever should be either. . . . I have dwelt longer on this subject, as I think you have, in some of your last letters, been somewhat deficient in that respect which your own good sense will at once convince you was, on all accounts, due, and which I know you feel the propriety of without any further observations."

On June 2, 1815, the father writes: —

"We have just received a letter from your uncle, James E. B. Finley, of Carolina. He fears you will remain in Europe, but hopes you have so much *amor patriæ* as to return and display your talents in raising the military and naval glory of the nation, by exhibiting on canvas some of her late naval and land actions, and also promote the fine arts among us. He is, you know, an enthusiastic Republican and patriot and a warm approver of the late war, but an amiable, excellent man. I am by no means certain that it would not be best for you to come home this fall and spend a year or two in this country in painting some portraits, but especially historical pieces and landscapes. You might, I think, in this way succeed in getting something to support you afterwards in Europe for a few years.

"I hope the time is not distant when artists in your profession, and of the first class, will be honorably patronized and supported in this country. In this case you can come and live with us, which would give us much satisfaction."

The young man still took a deep interest in affairs political, and speculated rather keenly on the outcome of the tremendous happenings on the Continent.

On June 26, 1815, he writes: —

"You will have heard of the dreadful battle in Flanders before this reaches you. The loss of the English is immense, indeed almost all their finest officers and the flower of their army; not less than 800 officers and upwards of 15,000 men, some say 20,000. But it has been decisive if the news of to-day be true, that *Napoleon has abdicated.* What the event of these unparalleled times will be no mortal can pretend to foresee. I have much to tell you when I see you. Perhaps you had better not write after the receipt of this, as it may be more than two months before an answer could be received.

"P.S. The papers of to-night confirm the news of this morning. Bonaparte is no longer a dangerous man; he has abdicated, and, in all probability, a republican form of government will be the future government of France, if they are capable of enjoying such a government. But no one can foresee events; there may be a long peace, or the world may be torn worse than it yet has been. Revolution seems to succeed revolution so rapidly that, in looking back on our lives, we seem to have lived a thousand years, and wonders of late seem to scorn to come alone; they come in clusters."

The battle in Flanders was the battle of Waterloo, which was fought on the 18th day of June, and on the 6th of July the allied armies again entered Paris. Referring to these events many years later, Morse said: —

"It was on one of my visits, in the year 1815, that an incident occurred which well illustrates the character of the great philanthropist [Mr. Wilberforce]. As I passed through Hyde Park on my way to Kensington Gore, I observed that great crowds had gathered, and rumors were rife that the allied armies had entered Paris, that Napoleon was a prisoner, and that the war was virtually at an end; and it was momentarily expected that the park guns would announce the good news to the people.

"On entering the drawing-room at Mr. Wilberforce's I found the company, consisting of Mr. Thornton [his memory must have played him false in this particular as Mr. Thornton died some time before], Mr. Macaulay,

Mr. Grant, the father, and his two sons Robert and Charles, and Robert Owen of Lanark, in quite excited conversation respecting the rumors that prevailed. Mr. Wilberforce expatiated largely on the prospects of a universal peace in consequence of the probable overthrow of Napoleon, whom naturally he considered the great disturber of the nations. At every period, however, he exclaimed: 'It is too good to be true, it cannot be true.' He was altogether skeptical in regard to the rumors.

"The general subject, however, was the absorbing topic at the dinner-table. After dinner the company joined the ladies in the drawing-room. I sat near a window which looked out in the direction of the distant park. Presently a flash and a distant dull report of a gun attracted my attention, but was unnoticed by the rest of the company. Another flash and report assured me that the park guns were firing, and at once I called Mr. Wilberforce's attention to the fact. Running to the window he threw it up in time to see the next flash and hear the report. Clasping his hands in silence, with the tears rolling down his cheeks, he stood for a few moments perfectly absorbed in thought, and, before uttering a word, embraced his wife and daughters, and shook hands with every one in the room. The scene was one not to be forgotten."

We learn from a letter of his mother's dated June 27, 1815, that the painting of the "Dying Hercules" had at last been received, but that the plaster cast of the same subject was still mysteriously missing. The painting was much admired, and the mother says: —

"Your friend Mr. Tisdale says the picture of the Hercules ought to be in Boston as the beginning of a gallery of paintings, and that the Bostonians ought not to permit it to go from here. Whether they will or not, I know not. I place no confidence in them, but they may take a fit into their heads to patronize the fine arts, and, in that case, they have it in their power undoubtedly to do as much as any city in this country towards their support."

Morse had now made up his mind to return home, although his parents, in their letters of that time, had given him leave to stay longer if he thought it would be for his best interest, but his father had made it clear that he must, from this time forth, depend on his own exertions. He hoped that (Providence permitting) he need only spend a year at home in earning enough money to warrant his returning to Europe. Providence, however, willed otherwise, and he did not return to Europe until fourteen years later.

The next letter is dated from Liverpool, August 8, 1815, and is but a short one. I shall quote the first few sentences: —

"I have arrived thus far on my way home. I left London the 5th and arrived in this place yesterday the 7th, at which time, within an hour, four years ago, I landed in England. I have not yet determined by what vessel to return; I have a choice of a great many. The Ceres is the first that sails, but I do not like her accommodations. The Liverpool packet sails about the 25th, and, as she has always been a favorite ship with me, it is not improbable I may return in her."

He decided to sail in the Ceres, however, to his sorrow, for the voyage home was a long and dreadful one. The record of those terrible fifty-eight days, carefully set down in his journal, reads like an Odyssey of misfortune and almost of disaster.

To us of the present day, who cross the ocean in a floating hotel, in a few days, arriving almost on the hour, the detailed account of the dangers, discomforts, and privations suffered by the travellers of an earlier period seems almost incredible. Brave, indeed, were our fathers who went down to the sea in ships, for they never knew when, if ever, they would reach the other shore, and there could be no C.Q.D. or S.O.S. flashed by wireless in the Morse code to summon assistance in case of disaster. In this case storm succeeded storm; head winds were encountered almost all the way across; fine weather and fair winds were the exception, and provisions and fresh water were almost exhausted.

The following quotations from the journal will give some idea of the terrors experienced by the young man, whose appointed time had not yet arrived. He still had work to do in the world which could be done by no other.

"*Monday, August 21, 1815.* After waiting fourteen days in Liverpool for a fair wind, we set sail at three o'clock in the afternoon with the wind at southeast, in company with upwards of two hundred sail of vessels, which formed a delightful prospect. We gradually lost sight of different vessels as it approached night, and at sunset they were dispersed all over the horizon. In the night the wind sprung up strong and fair, and in the morning we were past Holyhead.

"*Tuesday, 22d August.* Wind directly ahead; beating

all day; thick weather and gales of wind; passengers all sick and I not altogether well. Little progress to-day.

"*Wednesday, 23d August.* A very disagreeable day, boisterous, head winds and rainy. Beating across the channel from the Irish to the Welsh coast.

. . . .

"*Friday, 25th August.* Dreadful still; blowing harder and harder; quite a storm and a lee shore; breakers in sight, tacked and stood over again to the Irish shore under close-reefed topsails. At night saw Waterford light again.

. . . .

"*Monday, 28th August.* A fair wind springing up (ten o'clock). Going at the rate of seven knots on our true course. We have had just a week of the most disagreeable weather possible. I hope this is the beginning of better winds, and that, in reasonable time, we shall see our native shore.

"*Tuesday, 29th August.* Still disappointed in fair winds. . . . Since, then, I can find nothing consoling on deck, let us see what is in the cabin. All of us make six, four gentlemen and two ladies. Mrs. Phillips, Mrs. Drake, Captain Chamberlain, Mr. Bancroft, Mr. Lancaster, and myself. Our amusements are eating and drinking, sleeping and backgammon. Seasickness we have thrown overboard, and, all things considered, we try to enjoy ourselves and sometimes succeed.

. . . .

"*Thursday, 31st August.* Wind as directly ahead as it can blow; squally all night and tremendous sea. What a contrast does this voyage make with my first. This day makes the tenth day out and we have advanced towards home about three hundred miles. In my last voyage, on the tenth day, we had accomplished *one half* our voyage, sixteen hundred miles.

"*Friday, 1st September.* Dreadful weather; wind still ahead; foggy, rainy, and heavy swell; patience almost exhausted, but the will of Heaven be done. If this weather is to continue I hope we shall have fortitude to bear it. All is for the best.

. . . .

"*Saturday, 9th September.* Nineteenth day out and not yet more than one third of our way to Boston. Oh! when shall we end this tedious passage?

"*Sunday, 10th September.* Calm with dreadful sea. Early this morning discovered a large ship to the southward, dismasted, probably in the late gale. Discovered an unpleasant trait in our captain's character which I shall merely allude to. I am sorry to say he did not demonstrate that promptitude to assist a fellow creature in distress which I expected to find inherent in a seaman's breast, and especially in an American seaman's. It was not till after three or four hours' delay, and until the entreaties of his passengers and some threatening murmurs on my part of a public exposure in Boston of his conduct, that he ordered the ship to bear down upon the wreck, and then with slackened sail and much grumbling. A ship and a brig were astern of us, and, though farther by some miles from the distressed ship than we were, they instantly bore down for her, and rendered her this evening the assistance we might have done at noon. We are now standing on our way with a fair wind springing up at southeast, which I suppose will last a few hours. Spent the day in religious exercises, and was happy to observe on the part of the rest of the passengers a due regard for the solemnity of the day.

"*Monday, 11th September.* Wind still ahead and the sky threatening.—Ten o'clock. Beginning to blow hard; taking in sails one after another. — Three o'clock. A perfect storm; the gale a few days ago but a gentle breeze to it. . . . I never witnessed so tremendous a gale; the wind blowing so that it can scarcely be faced; the sea like ink excepting the whiteness of the surge, which is carried into the air like clouds of dust, or like the driving of snow. The wind piping through our bare rigging sounds most terrific; indeed, it is a most awful sight. The sea in mountains breaking over our bows, and a single wave dispersing in mist through the violence of the storm; ship rolling to such a degree that we are compelled to keep our berths; cabin dark with the deadlights in. Oh! who would go to sea when he can stay on shore! The wind in southwest driving us back again, so that we are losing all the advantages of our fair wind of yesterday, which lasted, as I supposed, two or three hours.

. . . .

"*Tuesday, 12th September.* Gale abated, but head wind still. . . .

"*Wednesday, 13th September.* All last night a tremendous storm from northwest.

"*Thursday, 14th September.* The storm increased to a tremendous height last night. The clouds at sunset were terrific in the extreme, and, in the evening, still more so with lightning. The sea has risen frightfully and everything wears a most alarming aspect. At 3 A.M. a squall struck us and laid us almost wholly under water; we came near losing our foremast. . . . None of us able to sleep from the dreadful noises; creakings and howlings and thousands of indescribable sounds. Lord! who can endure the terror of thy storm! . . . Yesterday's sea was as molehills to mountains compared with the sea to-day. . . .

"*Friday, 15th September.* The storm somewhat abated this morning, but still blowing hard from southwest. . . . Twenty-four days out to-day.

"*Saturday, 16th September.* Blowing a gale of wind from southwest. Noon almost calm for half an hour, when, on a sudden, the wind shifted to the northeast, when it blew such a hurricane that every one on board declared they never saw its equal. For four hours it blew so hard that all the sea was in a perfect foam, and resembled a severe snowstorm more than a dry blow. If the wind roared before, it now shrilly whistled through our rigging."

After some days of calm with winds sometimes favorable but light, and, when fresh, ahead, the journal continues: —

"*Monday, 25th September.* Another gale of wind last night, ahead, dreadful sea; took in sail and lay to all night. . . . Beginning to think of our provisions; bread mouldy and little left; sugar, little left; fresh provisions, little left; beans, none left; salt pork, little left; salt beef, a plenty; water, plenty; stores of passengers, some gone and the rest drawing to a conclusion; patience drawing to a conclusion; in short all is falling short and drawing to a conclusion except *our voyage* and *my journal*. . . .

"*Tuesday, 26th September.* . . . Find our captain to be a complete old woman; takes in sail at night and never knows when to set it again; the longer we know him, the more surly he grows; he is not even civil. . . . Several large turtles passed within a few feet of us yesterday and to-day, and, considering we are near the end of our provisions, one would have thought our captain would be anxious to take them; but no, it was too much trouble to lower the boat from the stern.

. . . .

"*Friday, 29th September.* Last night another dreadful gale, as severe as any since we have been out.

. . . .

"*Monday, 2d October.* Last night another gale of wind from northwest and is this morning still blowing hard and cold from the same quarter. What a dreadful passage is ours; we seem destined to have no fair wind, and to have a gale of wind every other day.

. . . .

"*Saturday, 7th October.* Wind still ahead and blowing hard; very cold and dismal. Oh! when shall we see home! . . . I thought I could observe a kind of warfare between the different winds since we have been at sea. The west wind seems to be the tyrant at present, as it were the Bonaparte of the air. He has been blowing his gales very lavishly, and no other wind has been able to check him with any success.

"I recollect on one day, while it was calm, a thick bank of clouds began to rise in the northeast; no other clouds were in the sky. They rose gently in the calm as if fearful of rousing their deadly foe in the west. Now they had gained one third of the heavens when, behold, in the southwest another bank of thick black clouds came rolling up, and, reddening in the rays of the setting sun, marched on, teeming with fury. They soon gained the middle of the heavens where the frightened northeast had not yet reached. They met, they mixed, the routed northeast skulked back, while the thick column of the southwest, having driven back its enemy, slowly returned to its repose, proudly displaying a thousand various colors, as if for victory.

"At another time success seemed to be more in favor of the northeast; for, shortly after this great defeat, the southwest came forth and, like a petty tyrant intoxicated with success, began to oppress the subject ocean. It blew its gales and filled the air with clouds and rain and fog. Suddenly the northeast, as under cover of the darkness, and as one driven to desperation, burst forth on its too confident enemy with redoubled fury. Old ocean groans at the dreadful conflict; for, as in the warring of two hostile armies on the domains of a neutral, the neutral suffers most severely, so the neutral ocean seemed doomed to bear the weight of all their rancor. The southwest flies affrighted. And now the northeast, vaunting forth, stalks with the rage of an angry demon over the waters; the ocean foams beneath his breath, it steams and smokes and heaves in agony its troubled bosom.

"But, alas! how few can bear prosperity; how few, when victory crowns their efforts, can rule with moderation; how often does it happen that we reënact the same scenes for which we punished our enemy. For now has the northeast become the tyrant and rules with tenfold rigor; he pours forth all his strength and, drunk with success as soldiers after a victory, at length sinks away into an inglorious calm.

"Now does the southwest collect his routed forces, checked but not conquered; he again advances on his recreant foe and seizes the vacant throne without a struggle. Ill-fated northeast! hadst thou but ruled with moderation when thou hadst gained, with masterly manœuvre, the throne of the air; hadst thou reserved thy forces against surprise, and not, with prodigal profuseness, lavished them on thy harmless subjects, thou hadst still been monarch of the sea and air; all would have blessed thee as the restorer of peace, and as the deliverer of the ocean from western despotism. But alas! how art thou fallen an everlasting example of overreaching oppression.

"This evening there is a fine fair wind from northeast carrying us on at the rate of five or six knots. This is the cause of the foregoing rhapsody. Had it been otherwise than a fair wind I should never have been in spirits to have written so much stuff."

Still tantalized by baffling head winds and alternating calms and gales, they were, however, gradually approaching the coast. Omitting the entries of the next eleven days, I shall quote the final pages of the journal.

"*Wednesday, 18th October.* Last night was a sleepless night to us all. Everything wore the appearance of a hard storm; all was dull in the cabin; scarce a word was spoken; every one wore a serious aspect and, as any one came from the deck into the cabin, the rest put up an inquisitive and apprehensive look, with now and then a faint, 'Well, how does it look now?' Our captain, as well as the passenger captain, were both alarmed, and

[FACSIMILE]

On board the Ship Ceres—
Boston Harbour—

My Dear Parents,

Thanks to a kind Providence who has preserved me through all dangers, I have at length arrived in my native land!— I send this just to prepare you, I shall be with you as soon as I can possibly get on shore.— we have had 58 days passage long, boisterous, and dangerous but more when I see you— Pray tell me by the bearer if I shall find all well.—

Yr Very affectionate Son

Saml. F. B. Morse

Oct. 18. 1815.

were poring over the chart in deep deliberation. A syllable was now and then caught from them, but all seemed despairing.

"At ten o'clock we lay to till twelve; at four again till five. Rainy, thick, and hazy, but not blowing very hard. All is dull and dismal; a dreadful state of suspense, between feelings of exquisite joy in the hope of soon seeing home, and feelings of gloomy apprehension that a few hours may doom us to destruction.

"*Half-past seven.* . . . Heaven be praised! The joyful tidings are just announced of *Land!!* Oh! who can conceive our feelings now? The wretch condemned to the scaffold, who receives, at the moment he expects to die, the joyful reprieve, he can best conceive the state of our minds.

"The land is Cape Cod, distant about ten miles. Joyful, joyful is the thought. To-night we shall, in all probability, be in Boston. We are going at the rate of seven knots.

"*Half-past 9.* Manomet land in sight.

"*Ten o'clock.* Cape Ann in sight.

"*Eleven o'clock.* Boston Light in sight.

"*One o'clock.* HOME!!!"

CHAPTER X

APRIL 10, 1816 — OCTOBER 5, 1818

Very little success at home. — Portrait of ex-President John Adams. — Letter to Allston on sale of his "Dead Man restored to Life." — Also apologizes for hasty temper. — Reassured by Allston. — Humorous letter from Leslie. — Goes to New Hampshire to paint portraits. — Concord. — Meets Miss Lucretia Walker. — Letters to his parents concerning her. — His parents reply. — Engaged to Miss Walker. — His parents approve. — Many portraits painted. — Miss Walker's parents consent. — Success in Portsmouth. — Morse and his brother invent a pump. — Highly endorsed by President Day and Eli Whitney. — Miss Walker visits Charlestown. — Morse's religious convictions. — More success in New Hampshire. — Winter in Charleston, South Carolina. — John A. Alston. — Success. — Returns north. — Letter from his uncle Dr. Finley. — Marriage.

THERE is no record of the meeting of the parents and the long-absent son, but it is easy to picture the joy of that occasion, and to imagine the many heart-to-heart conversations when all differences, political and otherwise, were smoothed over.

He remained at home that winter, but seems to have met with but slight success in his profession. His "Judgment of Jupiter" was much admired, but found no purchaser, nor did he receive any commissions for such large historical paintings as it was his ambition to produce. He was asked by a certain Mr. Joseph Delaplaine, of Philadelphia, to paint a portrait of ex-President John Adams for *half price*, the portrait to be engraved and included in "Delaplaine's Repository of the Lives and Portraits of Distinguished American Characters," and, from letters of a later date, I believe that Morse consented to this.

It appears that he must also have received but few, if any, orders for portraits, for, in the following summer, he started on a painting tour through New Hampshire, which proved to be of great moment to him in more ways than one.

Before we follow him on that tour, however, I shall quote from a letter written by him to his friend Washington Allston: —

BOSTON, April 10, 1816.

MY DEAR SIR, — I have but one moment to write you by a vessel which sails to-morrow morning. I wrote Leslie by New Packet some months since and am hourly expecting an answer.

I congratulate you, my dear sir, on the sale of your picture of the "Dead Man." I suppose you will have received notice, before this reaches you, that the Philadelphia Academy of Arts have purchased it for the sum of thirty-five hundred dollars. Bravo for our country!

I am sincerely rejoiced for you and for the disposition which it shows of future encouragement. I really think the time is not far distant when we shall be able to settle in our native land with profit as well as pleasure. Boston seems struggling in labor to bring forth an institution for the arts, but it will miscarry; I find it is all forced. They can talk, and talk, and say what a fine thing it would be, but nothing is done. I find by experience that what you have often observed to me with respect to settling in Boston is well founded. I think it will be the last in the arts, though, without doubt, it is capable of being the first, if the fit would only take them. Oh! how I miss you, my dear sir. I long to spend my evenings again with you and Leslie. I shall certainly visit Italy (should I live and no unforeseen event take place) in the course of a year or eighteen months. Could there not be some arrangement made to meet you and Leslie there?

He lived, but the "unforeseen event" occurred to make him alter all his plans. Further on in this same letter he says: —

"My conscience accuses me, and hardly too, of many instances of pettishness and ill-humor towards you, which make me almost hate myself that I could offend a temper like yours. I need not ask you to forgive it; I know you cannot harbor anger a minute, and perhaps have forgotten the instances; but I cannot forget them. If you had failings of the same kind and I could recollect any instances where you had spoken pettishly or ill-natured to me, our accounts would then have been balanced, they would have called for mutual forgetfulness and forgiveness; but when, on reflection, I find nothing of the kind to charge you with, my conscience severely upbraids me with ingratitude to you, to whom (under Heaven) I owe all the little knowledge of my art which I possess. But I hope still I shall prove grateful to you; at any rate, I feel my errors and must mend them."

Mr. Allston thus answers this frank appeal for forgiveness: —

My dear Sir, — I will not apologize for having so long delayed answering your kind letter, being, as you well know, privileged by my friends to be a lazy correspondent. I was sorry to find that you should have suffered the recollection of any hasty expressions you might have uttered to give you uneasiness. Be assured that they never were remembered by me a moment after, nor did they ever in the slightest degree diminish my regard or weaken my confidence in the sincerity of your friendship or the goodness of your heart. Besides, the consciousness of warmth in my own temper would have made me inexcusable had I suffered myself to dwell on an inadvertent word from another. I therefore beg you will no longer suffer any such unpleasant reflections to disturb your mind, but that you will rest assured of my unaltered and sincere esteem.

Your letter and one I had about the same time from my sister Mary brought the first intelligence of the sale of my picture, it being near three weeks later when I received the account from Philadelphia. When you recollect that I considered the "Dead Man" (from the untoward fate he had hitherto experienced) almost literally as a *caput mortuum*, you may easily believe that I was most agreeably surprised to hear of the sale. But, pleased as I was on account of the very seasonable pecuniary supply it would soon afford me, I must say that I was still more gratified at the encouragement it seemed to hold out for my return to America.

His friend Leslie, in a letter from London of May 7, 1816, writes: "Mr. West said your picture would have been more likely than any of them to obtain the prize had you remained."

In another letter from Leslie of September 6, 1816, occurs this amusing passage: —

"The *Catalogue Raisonné* appeared according to promise, but is not near so good as the one last year. At the conclusion the author says that Mr. Payne Knight told the directors it was the custom of the Greek nobility to strip and exhibit themselves naked to the artists in various attitudes, that they might have an opportunity of studying fine form. Accordingly those public-spirited men, the directors, have determined to adopt the plan, and are all practising like mad to prepare themselves for the ensuing exhibition, when they are to be placed on pedestals.

"It is supposed that Sir G. Beaumont, Mr. Long, Mr. Knight, etc., will occupy the principal lights. The Marquis of Stafford, unfortunately, could not recollect the attitude of any one antique figure, but was found practising having the head of the Dying Gladiator, the body of the Hercules, one leg of the Apollo, and the other of the Dancing Faun, turned the wrong way. Lord Mulgrave, having a small head, thought of representing the Torso, but he did not know what to do with his legs, and was afraid that, as Master of the Ordnance, he could not dispense with his *arms*."

In the beginning of August, 1816, the young man started out on his quest for money. This was frankly the object of his journey, but it was characteristic of his buoyant and yet conscientious nature that, having once made up his mind to give up, for the present, all thoughts of pursuing the higher branches of his art, he took up with zest the painting of portraits.

So far from degrading his art by pursuing a branch of it which he held to be inferior, he still, by conscientious work, by putting the best of himself into it, raised it to a very high plane; for many of his portraits are now held by competent critics to rank high in the annals of art, by some being placed on a level with those of Gilbert Stuart.

On August 8, 1816, he writes to his parents from Concord, New Hampshire: —

"I have been in this place since Monday evening. I arrived safely. . . . Massabesek Pond is very beautiful, though seen on a dull day. I think that one or two elegant views might be made from it, and I think I must sketch it at some future period.

"I have as yet met with no success in portraits, but hope, by perseverance, I shall be able to find some. My stay in this place depends on that circumstance. If none offer, I shall go for Hanover on Saturday morning.

"The scenery is very fine on the Merrimack; many fine pictures could be made here alone. I made a little sketch near Contoocook Falls yesterday. I go this morning with Dr. McFarland to see some views. Colonel Kent's family are very polite to me, and I never felt in better spirits; the weather is now fine and I feel as though I was growing fat."

CONCORD, August 16, 1816.

I am still here and am passing my time very agreeably. I have painted five portraits at fifteen dollars each and have two more engaged and many more talked of. I think I shall get along well. I believe I could make an independent fortune in a few years if I devoted myself exclusively to portraits, so great is the desire for good portraits in the different country towns.

He must have been a very rapid worker to have painted five portraits in eight days; but, perhaps, on account of the very modest price he received, these were more in the nature of quick sketches.

The next letter is rather startling when we recall his recent assertions concerning "Mrs. Love" and the joys of a bachelor existence.

CONCORD, August 20, 1816.

MY DEAR PARENTS, — I write you a few lines just to say I am well and very industrious. Next day after to-morrow I shall have received one hundred dollars, which I think is pretty well for three weeks. I shall probably stay here a fortnight from yesterday.

I have other attractions besides money in this place. Do you know the Walkers of this place? Charles Walker Esq., son of Judge Walker, has two daughters, the elder, very beautiful, amiable, and of an excellent disposition. This is her character in town. I have enquired particularly of Dr. McFarland respecting the family, and his answer is every way satisfactory, except that they are not professors of religion. He is a man of family and great wealth. This last, you know, I never made a principal object, but it is somewhat satisfactory to know that in my profession.

I may flatter myself, but I think I might be a successful suitor.

You will, perhaps, think me a terrible harum-scarum fellow to be continually falling in love in this way, but I have a dread of being an old bachelor, and I am now twenty-five years of age.

There is still no need of hurry; the young lady is but sixteen. But all this is thinking aloud to you; I make you my confidants; I wish your advice; nothing shall be done precipitately.

Of course all that I say is between you and me, for it

all may come to nothing; I have *some experience* that way.

What I have done I have done prayerfully. I have prayed to the Giver of every good gift that He will direct me in this business; that, if it will not be to his glory and the good of his Kingdom, He will frustrate all; that, if He grants me prosperity, He will grant me a heart to use it aright; and, if adversity, that He will teach me submission to his will; and that, whatever may be my lot here, I may not fall short of eternal happiness hereafter.

I hope you will remember me in your prayers, and especially in reference to a connection in life.

I do not think that his parents took this matter very seriously at first. His was an intensely affectionate nature, and they had often heard these same raptures before. However, like wise parents, they did not scoff. His mother wrote on August 23, 1816, in answer: "With respect to the other confidential matter, I hope the Lord will direct you to a proper choice. We know nothing of the family, good or bad. We do not wish you to be an old bachelor, nor do we wish you to precipitate yourself and others into difficulties which you cannot get rid of."

In the same letter his father says: "In regard to the subject on which you ask our advice, we refer it, after the experience you have had, and with the advice you have often had from us, to your own judgment. Be not hasty in entering into any engagement; enquire with caution and delicacy; do everything that is honorable and gentlemanly respecting yourself and those concerned. 'Pause, ponder, sift. — Judge before friendship — then confide till death.' (Young.) Above all, commit the subject to God in prayer and ask his guidance and blessing. I am glad to find you are doing this."

How well he obeyed his father's injunctions may be gathered from the following letter, which speaks for itself: —

CONCORD, September 2, 1816.

MY DEAR PARENTS, — I have just received yours of August 29. I leave town to-morrow morning, probably for Hanover, as there is no conveyance direct to Walpole.

I have had no more portraits since I wrote you, so that I have received just one hundred dollars in Concord. The last I took for ten dollars, as the person I painted obtained four of my sitters for me. . . .

With respect to the confidential affair, everything is successful beyond my most sanguine expectations. The more I know of her the more amiable she appears. She is very beautiful and yet no coquetry; she is modest, quite to diffidence, and yet frank and open-hearted. Wherever I have enquired concerning her I have invariably heard the same character of — "remarkably amiable, modest, and of a sweet disposition." When you learn that this is the case I think you will not accuse me of being hasty in bringing the affair to a crisis. I ventured to tell her my whole heart, and instead of obscure and ambiguous answers, which some would have given to tantalize and pain one, she frankly, but modestly and timidly, told me it was mutual. Suffice it to say we are *engaged.*

If I know my parents I know they will be pleased with

this amiable girl. Unless I was confident of it, I should never have been so hasty. I have not yet mentioned it to her parents; she requested me to defer it till next summer, or till I see her again, lest she should be thought hasty. She is but sixteen and is willing to wait two or three years if it is for our mutual interest.

Never, never was a human being so blest as I am, and yet what an ungrateful wretch I have been. Pray for me that I may have a grateful heart, for I deserve nothing but adversity, and yet have the most unbounded prosperity.

The father replies to this characteristic letter on September 4, 1816: —

"I have just received yours of the 2d inst. Its contents were deeply interesting to us, as you will readily suppose. It accounts to us why you have made so long a stay at Concord. . . . So far as we can judge from your representations (which are all we have to judge from), we cannot refuse you our approbation, and we hope that the course, on which you have entered with your characteristic rapidity and decision, will be pursued and issue in a manner which will conduce to the happiness of all concerned. . . .

"We think *her* parents should be made acquainted with the state of the business, as she is so young and the thing so important to them."

The son answers this letter, from Walpole, New Hampshire, on September 7, 1816, thus naïvely: "You think the parents of the young lady should be made acquainted with the state of the business. I feel some degree of awkwardness as it respects that part of the affair; I don't know the manner in which it ought to be done. I wish you would have the goodness to write me immediately (at Walpole, to care of Thomas Bellows, Esq.) and inform me what I should say. Might I communicate the information by writing?"

Here he gives a detailed account of the family, and, for the first time, mentions the young lady's name — Lucretia Pickering Walker — and continues: —

"You ask how the family have treated me. They are all aware of the attachment between us, for I have made my attention so open and so marked that they all must have perceived it. I know that Lucretia must have had some conversation with her mother on the subject, for she told me one day, when I asked her what her mother thought of my constant visits, that her mother said she 'did n't think I cared much about her,' in a pleasant way. All the family have been extremely polite and attentive to me; I received constant invitations to dinner and tea, indeed every encouragement was given me. . . .

"I painted two hasty sketches of scenery in Concord. I meet with no success in Walpole. *Quacks* have been before me."

There is always a touch of quaint, dry humor in his mother's letters in spite of their great seriousness, as witness the following extracts from a letter of September 9, 1816: —

"We hope you will feel more than ever the absolute necessity laid upon you to procure for yourself and those you love a maintenance, as neither of you can subsist long upon air. . . . Remember it takes a great many hundred dollars to *make* and to *keep* the pot a-boiling.

"I wish to see the young lady who has captivated you

so much. I hope she loves religion, and that, if you and she form a connection for life, some *five or six years hence*, you may go hand in hand to that better world where they neither marry nor are given in marriage. . . .

"You have not given us any satisfaction in respect to many things about the young lady which you ought to suppose we should be anxious to know. All you have told us is that she is handsome and amiable. These are good as far as they go, but there are a great many etcs., etcs., that we want to know.

"Is she acquainted with domestic affairs? Does she respect and love religion? How many brothers and sisters has she? How old are they? Is she healthy? How old are her parents? What will they be likely to do for her some years hence, say when she is twenty years old?

"In your next answer at least some of these questions. You see your mother has not lived twenty-seven years in New England without learning to ask questions."

These questions he had already answered in a letter which must have crossed his mother's.

On September 23, 1816, he writes from Windsor, Vermont: —

"I am still here but shall probably leave in a week or two. I long to get home, or, at least, as far on my way as *Concord*. I think I shall be tempted to stay a week or two there. . . . I do not like Windsor very much. It is a very dissipated place, and dissipation, too, of the lowest sort. There is very little gentleman's society."

WINDSOR, VERMONT, September 28, 1816.

I am still in this place. . . . I have written Lucretia on the subject of acquainting her parents, and I have no doubt she will assent. . . . I hear her spoken of in this part of the country as very celebrated, both for her beauty and, particularly, for her disposition; and this I have heard without there being the slightest suspicion of any attachment, or even acquaintance, between us. This augurs well most certainly. I know she is considered in Concord as the first girl in the place. (You know I always aimed highest.) The more I think of this attachment the more I think I shall not regret the *haste* (if it may be so called) of this proposed connection. . . .

I am doing pretty well in this place, better than I expected; I have one more portrait to do before I leave it. . . . I should have business, I presume, to last me some weeks if I could stay, but I long to get home *through Concord*. . . .

Mama's scheme of painting a large landscape and selling it to General Bradley for two hundred dollars, must give place to another which has just come into my head: that of sending to you for my great canvas and painting the quarrel at Dartmouth College, as large as life, with all the portraits of the trustees, overseers, officers of college, and students; and, if I finish it next week, to ask five thousand dollars for it and then come home in a coach and six and put Ned to the blush with his nineteen subscribers a day. Only think, $5000 a week is $260,000 a year, and, if I live ten years, I shall be worth $2,600,000; a very pretty fortune for this time of day. Is it not a grand scheme?

The remark concerning his brother Sidney Edwards's subscribers refers to a religious newspaper, the "Boston Recorder," founded and edited by him. It was one of the first of the many religious journals which, since that time, have multiplied all over the country.

Continuing his modestly successful progress, he writes next from Hanover, on October 3, 1816:—

"I arrived in this place on Tuesday evening and am painting away with all my might. I am painting Judge Woodward and lady, and think I shall have many more engaged than I can do. I painted seven portraits at Windsor, one for my board and lodging at the inn, and one for ten dollars, very small, to be sent in a letter to a great distance; so that in all I received eighty-five dollars in money. I have five more engaged at Windsor for next summer. So you see I have not been idle.

"I *must* spend a fortnight at Concord, so that I shall not probably be at home till early in November.

"I think, with proper management, that I have but little to fear as to this world. I think I can, with industry, average from two to three thousand dollars a year, which is a tolerable income, though *not equal to* $2,600,000!"

CONCORD, October 14, 1816.

I arrived here on Friday evening in good health and spirits from Hanover. I painted four portraits altogether in Hanover, and have many engaged for next summer. I presume I shall paint some here, though I am uncertain.

I found Lucretia in good health, very glad to see me. She improves on acquaintance; she is, indeed, a most amiable, affectionate girl; I know you will love her. She has consented that I should inform her parents of our attachment. I have, accordingly, just sent a letter to her father (twelve o'clock), and am now in a state of suspense anxiously waiting his answer. Before I close this, I hope to give you the result.

Five o'clock. I have just called and had a conversation (by request) with Mr. Walker, and I have the satisfaction to say: "I have Lucretia's parents' entire approbation." Everything successful! Praise be to the giver of every good gift! What, indeed, shall I render to Him for all his unmerited and continually increasing mercies and blessings?

In a letter to Miss Walker from a girl friend we find the following: —

"You appear to think, dear Lucretia, that I am possessed of quite an insensible *heart;* pardon me if I say the same of you, for I have heard that several have become candidates for your affections, but that you remained unmoved until Mr. M., of Charlestown, made his appearance, when, I understand, you did hope that his sentiments in your favor were reciprocal.

"I rejoice to hear this, for, though I am unacquainted with that gentleman, yet, when I heard he was likely to become a successful suitor, I have made some enquiries concerning him, and find he is possessed of every excellent and amiable quality that I should wish the person to have who was to become the husband of so dear a friend as yourself."

Morse must have returned home about the end of October, for we find no more letters until the 14th of December, when he writes from Portsmouth, New Hampshire: —

"I should have written you sooner but I have been employed in settling myself. I thought it best not to be precipitate in fixing on a place to board and lodge, but first to sound the public as to my success. Every one thinks I shall meet with encouragement, and, on the strength of this, I have taken lodgings and a room at Mrs. Ringe's in Jaffrey Street; a very excellent and central situation. . . . I shall commence on Monday morning with Governor Langdon's portrait. He is very kind and attentive to me, as, indeed, are all here, and will do everything to aid me. I wish not to raise high expectations, but I think I shall succeed tolerably well."

About this time Finley Morse and his brother Edwards had jointly devised and patented a new "flexible piston-pump," from which they hoped great things. Edwards, always more or less of a wag, proposed to call it "Morse's Patent Metallic Double-headed Ocean-Drinker and Deluge-Spouter Valve Pump-Boxes."

It was to be used in connection with fire-engines, and seems really to have been an excellent invention, for President Jeremiah Day, of Yale College, gave the young inventors his written endorsement, and Eli Whitney, the inventor of the cotton-gin, thus recommends it: "Having examined the model of a fire-engine invented by Mr. Morse, with pistons of a new construction, I am of opinion that an engine may be made on that principle (being more simple and much less expensive), which would have a preference to those in common use."

In the letters of the year 1817 and of several following years, even in the letters of the young man to his *fiancée*, many long references are made to this pump and to the varying success in introducing it into general use. I shall not, however, refer to it again, and only mention it to show the bent of Morse's mind towards invention.

He spent some time in the early part of 1817 in Portsmouth, New Hampshire, meeting with success in his profession. Miss Walker was also there visiting friends, so we may presume that his stay was pleasant as well as profitable.

In February of that year he accompanied his *fiancée* to Charlestown, his parents, naturally, wishing to make the acquaintance of the young lady, and then returned to Portsmouth to finish his work there.

The visit of Miss Walker to Charlestown gave great satisfaction to all concerned. On March 4, 1817, Morse writes to his parents from Portsmouth: "I am under the agreeable necessity (shall I say) of postponing my return . . . in consequence of a *press of business*. I shall have three begun to-night; one sat yesterday (a large one), and two will sit to-day (small), and three more have it in serious contemplation. This unexpected occurrence will deprive me of the pleasure of seeing you this week at least."

And on the next day, March 5, he writes: "The unexpected application of three sitters at a time completely stopped me. Since I wrote I have taken a first sitting of a fourth (large), and a fifth (large) sits on Friday morning; so you see I am over head and ears in business."

As it is necessary to a clear understanding of Morse's character to realize the depth of his religious convictions, I shall quote the following from this same letter of March 5: —

"I wish much to know the progress of the Revival, how many are admitted next communion, and any religious news.

"I have been in the house almost ever since I came

from home sifting the scheme of Universal Salvation to the bottom. What occasioned this was an occurrence on the evening of Sunday before last. I heard the bell ring for lecture and concluded it was at Mr. Putnam's; I accordingly sallied out to go to it, when I found that it was in the Universalist meeting-house.

"As I was out and never in a Universalist meeting, I thought, for mere curiosity, I would go in. I went into a very large meeting-house; the meeting was overflowing with people of both sexes, and the singing the finest I have heard in Portsmouth. I was struck with the contrast it made to Mr. Putnam's sacramental lecture; fifteen or sixteen persons thinly scattered over the house, and the choir consisting of four or five whose united voice could scarcely be heard in the farthest corner of the church, and, when heard, so out of harmony as to set one's teeth on edge.

"The reflections which this melancholy contrast caused I could not help communicating to Mr. Putnam in the words of Mr. Spring's sermon, '*something must be done.*' He agreed it was a dreadful state of society here but almost gave up as hopeless. I told him he never should yield a post like this to the Devil without a struggle; and, at any rate, I told him that the few Christians that there were (and, indeed, they are but as one to one thousand) could pray, and I thought it was high time. I told him I would do all in my power to assist him in any scheme where I could be of use."

The year 1817 was spent by the young man in executing the commissions which had been promised him the year before in New Hampshire. In all his journeyings back and forth the road invariably led through Concord, and the pure love of the young people for each other increased as the months rolled by. I shall not profane the sacredness of this love by introducing any of the more intimate passages of their letters of this and of later years. The young girl responded readily to the religious exhortations of her *fiancé* and became a sincere and devout Christian.

It will not be necessary to follow him in this journey, as the experiences were but a repetition of those of the year before. He painted many portraits in Concord, Hanover, and other places, and finally concluded to venture on a trip to Charleston, South Carolina, where his kinsman, Dr. Finley, and Mr. John A. Alston had urged him to come, assuring him good business.

On January 27, 1818, he arrived in that beautiful Southern city and thus announced his arrival to his parents: "I find myself in a new climate, the weather warm as our May. I have been introduced to a number of friends. I think my prospects are favorable."

At first, however, the promised success did not materialize, and it was not until after many weeks of waiting that the tide turned. But it did turn, for an excellent portrait of Dr. Finley, one of the best ever painted by Morse, aroused the enthusiasm of the Charlestonians, and orders began to pour in, so that in a few weeks he was engaged to paint one hundred and fifty portraits at sixty dollars each. Quite an advance over the meagre fifteen dollars he had received in New England. But for some of his more elaborate productions he received even more, as the following extract from a letter of Mr. John A. Alston, dated April 7, 1818, will prove: —

"I have just received your favor of the 30th ultimo,

and thank you very cordially for your goodness in consenting to take my daughter's full-length likeness in the manner I described, say twenty-four inches in length. I will pay you most willingly the two hundred dollars you require for it, and will consider myself a gainer by the bargain. I shall expect you to decorate this picture with the most superb landscape you are capable of designing, and that you will produce a masterpiece of painting. I agree to your taking it with you to the northward to finish it. Be pleased to represent my daughter in the finest attitude you can conceive."

Mr. Alston was a generous patron and paid the young artist liberally for the portraits of his children. In recognition of this Morse presented him with his most ambitious painting, "The Judgment of Jupiter." Mr. Alston prized this picture highly during his lifetime, but after his death it was sold and for many years was lost sight of. It was purchased long afterwards in England by an American gentleman, who, not knowing who the painter was, gave it to a niece of Morse's, Mrs. Parmalee, and it is still, I believe, in the possession of the family.

While he was in Charleston his father wrote to him of the dangerous illness of his mother with what he called a "peripneumony," which, from the description, must have been the term used in those days for pneumonia. Her life was spared, however, and she lived for many years after this.

In June of the year 1818, Morse returned to the North and spent the summer in completing such portraits as he had carried with him in an unfinished state, and in painting such others as he could procure commissions for. He planned to return to Charleston in the following year, but this time with a young wife to accompany him.

His uncle, Dr. Finley, writing to him on June 16, says: —

"Your letter of 2d instant, conveying the pleasing intelligence of your safe and very short passage and happy meeting with your affectionate parents at your own home, came safe to hand in due time. . . . And so Lucretia was expected and you intended to surprise her by your unlooked-for presence.

"Finley, I am afraid you will be too happy. You ought to meet a little rub or two or you will be too much in the clouds and forget that you are among mortals. Let me see if I cannot give you a friendly twist downwards.

"Your pictures — aye — suppose I should speak of them and what is said of them during your absence. I will perform the office of him who was placed near the triumphal car of the conqueror to abuse him lest he should be too elated.

"Well — 'His pictures,' say people, 'are undoubtedly good likenesses, but he paints carelessly and in too much haste and his draperies are not well done. He must be more attentive or he will lose his reputation.' 'See,' say others, 'how he flatters.' 'Oh!' says another, 'he has not flattered me'; etc., etc.

"By the bye, I saw old General C. C. Pinckney yesterday, and he told me, in his laughing, humorous way, that he had requested you to draw his brother Thomas twenty years younger than he really was, so as to be a companion to his own when he was twenty years younger than at this time, and to flatter him as he had directed Stuart to do so to him."

Morse had now abandoned his idea of soon returning to Europe; he renounced, for the present, his ambition to devote himself to the painting of great historical pictures, and threw himself with enthusiasm into the painting of portraits. He had an added incentive, for he wished to marry at once, and his parents and those of his *fiancée* agreed that it would be wise for the young people to make the venture. Everything seemed to presage success in life, at least in a modest way, to the young couple.

On the 6th of October, 1818, the following notice appeared in the New Hampshire "Patriot," of Concord: "Married in this town, October 1st, by Rev. Dr. McFarland, Mr. Samuel F. B. Morse (the celebrated painter) to Miss Lucretia Walker, daughter of Charles Walker, Esq."

On the 5th of October the young man writes to his parents: —

"I was married, as I wrote you I should be, on Tuesday morning last. We set out at nine o'clock and reached Amherst over bad roads at night. The next day we continued our journey through Wilton to New Ipswich, eighteen miles over one of the worst roads I ever travelled, all uphill and down and very rocky, and no tavern on the road. We enquired at New Ipswich our best route to Northampton, where we intended to go to meet Mr. and Mrs. Cornelius, but we found on enquiry that there were nothing but cross-roads and these very bad, and no taverns where we could be comfortably accommodated. Our horse also was tired, so we thought our best way was to return. Accordingly the next day we started for Concord, and arrived on Friday evening safe home again.

"Lucretia wishes to spend this week with her friends, so that I shall return (Providence permitting) on this day week, and reach home by Tuesday noon, probably to dinner. We are both well and send a great deal of love to you all. Mr. and Mrs. Walker wish me to present their best respects to you. We had delightful weather for travelling, and got home just in season to escape Saturday's rain."

CHAPTER XI

NOVEMBER 19, 1818 — MARCH 31, 1821.

Morse and his wife go to Charleston, South Carolina. — Hospitably entertained and many portraits painted. — Congratulates Allston on his election to the Royal Academy. — Receives commission to paint President Monroe. — Trouble in the parish at Charlestown. — Morse urges his parents to leave and come to Charleston. — Letters of John A. Alston. — Return to the North. — Birth of his first child. — Dr. Morse and his family decide to move to New Haven. — Morse goes to Washington. — Paints the President under difficulties. — Hospitalities. — Death of his grandfather. — Dr. Morse appointed Indian Commissioner. — Marriage of Morse's future mother-in-law. — Charleston again. — Continued success. — Letters to Mrs. Ball. — Liberality of Mr. Alston. — Spends the summer in New Haven. — Returns to Charleston, but meets with poor success. — Assists in founding Academy of Arts, which has but a short life. — Goes North again.

THE young couple decided to spend the winter in Charleston, South Carolina, where Morse had won a reputation the previous winter as an excellent portrait-painter, and where much good business awaited him.

The following letter was written to his parents: —

SCHOONER TONTINE, AT ANCHOR OFF CHARLESTON LIGHTHOUSE,
THURSDAY, November 19, 1818, 5 o'clock P.M.

We have arrived thus far on our voyage safely through the kind protection of Providence. We have had a very rough passage attended with many dangers and more fears, but have graciously been delivered from them all. It is seven days since we left New York. If you recollect that was the time of my last passage in this same vessel. She is an excellent vessel and has the best captain and accommodations in the trade.

Lucretia was a little seasick in the roughest times, but, on the whole, bore the voyage extremely well. She seems a little downcast this afternoon in consequence of feeling as if she was going among strangers, but I tell her she will overcome it in ten minutes' interview with Uncle and Aunt Finley and family.

She is otherwise very well and sends a great deal of love to you all. Please let Mr. and Mrs. Walker know of our arrival as soon as may be. I will leave the remainder of this until I get up to town. We hope to go up when the tide changes in about an hour.

FRIDAY MORNING, 20th,
AT UNCLE FINLEY'S.

We are safely housed under the hospitable roof of Uncle Finley, where they received us, as you might expect, with open arms. He has provided lodgings for us at ten dollars per week. I have not yet seen them; shall go directly.

I received a letter from Richard at Savannah; he writes in fine spirits and feels quite delighted with the hospitable people of the South.

This refers to his brother Richard Carey Morse, who was still pursuing his theological studies.

The visit of the young couple to Charleston was a most enjoyable one, and the artist found many patrons eager to be immortalized by his brush.

On December 22, 1818, he writes to his parents: —

"Lucretia is well and contented. She makes many friends and we receive as much attention from the hospitable Carolinians as we can possibly attend to. She is esteemed quite handsome here; she has grown quite fleshy and healthy, and we are as happy in each other as you can possibly wish us.

"There are several painters arrived from New York, but I fear no competition; I have as much as I can do."

As a chronicle of fair weather, favorable winds, and blue skies is apt to grow monotonous, I shall pass rapidly over the next few years, only selecting from the voluminous correspondence of that period a few extracts which have more than a passing interest.

On February 4, 1819, he writes to his friend and master, Washington Allston, who had now returned to Boston: —

"Excuse my neglect in not having written you before this according to my promise before I left Boston. I can only plead as apology (what I know will gratify you) a multiplicity of business. I am painting from morning till night and have continual applications. I have added to my list, this season only, to the amount of three thousand dollars; that is since I left you. Among them are three full lengths to be finished at the North, I hope in Boston, where I shall once more enjoy your criticisms.

"I am exerting my utmost to improve; every picture I try to make my best, and in the evening I draw two hours from the antique as I did in London; for I ought to inform you that I fortunately found a fine 'Venus de Medicis' without a blemish, imported from Paris sometime since by a gentleman of this city who wished to dispose of it; also a young Apollo which was so broken that he gave it to me, saying it was useless. I have, however, after a great deal of trouble, put it together entirely, and these two figures, with some fragments, — hands, feet, etc., — make a good academy. Mr. Fraser, Mr. Cogdell, Mr. Fisher, of Boston, and myself meet here of an evening to improve ourselves. I feel as much enthusiasm as ever in my art and love it more than ever. A few years, at the rate I am now going on, will place me independent of public patronage.

"Thus much for myself, for you told me in one of your letters from London that I must be more of an egotist or you should be less of one in your letters to me, which I should greatly regret.

"And now, permit me, my dear sir, to congratulate you on your election to the Royal Academy. I know you will believe me when I say I jumped for joy when I heard it. Though it cannot add to your merit, yet it will extend the knowledge of it, especially in our own country, where we are still influenced by foreign opinion, and more justly, perhaps, in regard to taste in the fine arts than in any other thing."

On March 1, 1819, the Common Council of Charleston passed the following resolution: —

"Resolved unanimously that His Honor the Intendant be requested to solicit James Monroe, President of the United States, to permit a full-length likeness to be taken for the City of Charleston, and that Mr. Morse be requested to take all necessary measures for executing the said likeness on the visit of the President to this city.

"Resolved unanimously that the sum of seven hundred and fifty dollars be appropriated for this purpose.

"Extract from the minutes.

"WILLIAM ROACH, JR.,
"Clerk of Council."

This portrait of President Monroe was completed later on and still hangs in the City Hall of Charleston. I shall have occasion to refer to it again.

Morse, in a letter to his parents of March 26, 1819, says: —

"Two of your letters have been lately received detailing the state of the parish and church. I cannot say I was surprised, for it is what might be expected from Charlestown people. . . . As to returning home in the way I mentioned mama need not be at all uneasy on that score. It is necessary I should visit Washington, as the President will stay so short a time here that I cannot complete the head unless I see him in Washington. . . . Now as to the parish and church business, I hope all things will turn out right yet, and I can't help wishing that nothing may occur to keep you any longer in that nest of vipers and conspirators. I think with Edwards decidedly that, on mama's account alone, you should leave a place which is full of the most unpleasant associations to all the family, and retire to some place of quiet to enjoy your old age.

"Why not come to Charleston? Here is a fine place for usefulness, a pleasant climate especially for persons advanced in life, and your children here; for I think seriously of settling in Charleston. Lucretia is willing, and I think it will be much for my advantage to remain through the year. Richard can find a place here if he will, and Edwards can come on and be *Bishop* or *President* or *Professor* in some of the colleges (for I can't think of him in a less character) after he has graduated.

"I wish seriously you would think of this. Your friends here would greatly rejoice and an opening could be found, I have no doubt. Christians want their hands strengthened, and a veteran soldier, like papa, might be of great service here in the infancy of the *Unitarian Hydra*, who finds a population too well adapted to receive and cherish its easy and fascinating tenets."

All this refers to a movement organized by the enemies of Dr. Morse to oust him from his parish in Charlestown. He was a militant fighter for orthodoxy and an uncompromising foe to Unitarianism, which was gradually obtaining the ascendancy in and near Boston. The movement was finally successful, as we shall see later, but they did not go as far from their old haunts as Charleston.

I shall not attempt to argue the rights and wrongs of the case, which seem to have been rather complicated, for Dr. Morse, more than a year after this, in writing to a friend says: "The events of the last fifteen months are still involved in impenetrable mystery, which I doubt not will be unravelled in due time."

The winter and spring of 1819 were spent by the young couple both pleasantly and profitably in Charleston. The best society of that charming city opened its arms to them and orders flowed in in a steady stream. Mr. John A. Alston was a most generous patron, ordering many portraits of his children and friends, and sometimes insisting on paying the young man even more than the price agreed upon.

In a letter to Morse he says: "Which of my friends was it who lately observed to you that I had a picture mania? You made, I understand, a most excellent reply, 'You wished I would come to town, then, and bite a dozen.' Indeed, my very good sir, was it in my power to excite in them a just admiration of your talents, I would readily come to town and bite the whole community."

And in another letter of April 10, 1819, Mr. Alston says: "Your portrait of my daughter was left in Georgetown [South Carolina], at the house of a friend; nearly all of the citizens have seen it, and I really think it will occasion you some applications. . . . Every one thought himself at liberty to make remarks. Some declared it to be a good likeness, while others insisted it was not so, and several who made such remarks, I *knew* had *never* seen my daughter. At last a rich Jew gentleman observed, 'it was the *richest* piece of painting he had ever seen.' This being so much in character that I assure you, sir, I could contain myself no longer, which, spreading among the audience, occasioned not an unpleasant moment."

Morse and his young wife returned to the North in the early summer of 1819, and spent the summer and fall with his parents in Charlestown. The young man occupied himself with the completion of the portraits which he had brought with him from the South, and his wife was busied with preparations for the event which is thus recorded in a letter of Dr. Morse's to his son Sidney Edwards at Andover: "Since I have been writing the above, Lucretia has presented us with a fine granddaughter and is doing well. The event has filled us with joy and gratitude."

The child was christened Susan Walker Morse. In the mean time the distressing news had come from Charleston of the sudden death of Dr. Finley, to whose kindly affection and influence Morse owed much of the pleasure and success of his several visits to Charleston.

Affairs had come to a crisis in the parish at Charlestown, and Dr. Morse decided to resign and planned to move to New Haven, Connecticut, with his family in the following spring.

The necessity for pursuing his profession in the most profitable field compelled Morse to return to Charleston by way of Washington in November, and this time he had to go alone, much against his inclinations.

He writes to his mother from New York on November 28, 1819: "I miss Lucretia and little Susan more than you can think, and I shall long to have us all together at New Haven in the spring."

His object in going to Washington was to paint the portrait of the President, and of this he says in a letter: "I began on Monday to paint the President and have almost completed the head. I am thus far pleased with it, but I find it very perplexing, for he cannot sit more than ten or twenty minutes at a time, so that the moment I feel engaged he is called away again. I set my palette to-day at ten o'clock and waited until four o'clock this afternoon before he came in. He then sat ten minutes and we were called to dinner. Is not this trying to one's patience?"

"*December 17, 1819.* I have been here nearly a fortnight. I commenced the President's portrait on Monday and shall finish it to-morrow. I have succeeded to my satisfaction, and, what is better, to the satisfaction of himself and family; so much so that one of his daughters wishes me to copy the head for her. They all say that mine is the best that has been taken of him. The daughter told me (she said as a secret) that her father was delighted with it, and said it was the only one that in his opinion looked like him; and this, too, with Stuart's in the room.

"The President has been very kind and hospitable to me; I have dined with him three times and taken tea as often; he and his family have been very sociable and unreserved. I have painted him at his house, next room to his cabinet, so that when he had a moment to spare he would come in to me.

"Wednesday evening Mrs. Monroe held a drawing-room. I attended and made my bow. She was splendidly and tastily dressed. The drawing-room and suite of rooms at the President's are furnished and decorated in the most splendid manner; some think too much so, but I do not. Something of splendor is certainly proper about the Chief Magistrate for the credit of the nation. Plainness can be carried to an extreme, and in national buildings and establishments it will, with good reason, be styled meanness."

"*December 23, 1819.* It is obviously for my interest to hasten to Charleston, as I shall there be immediately at work, and this is the more necessary as there is a fresh gang of adventurers in the brush line gone to Charleston before me."

A short while after this he received the news of the death of his grandfather, Jedediah Morse, at Woodstock, Connecticut, on December 29, aged ninety-four years. Mr. Prime says of him: "He was a strong man in body and mind, an able and upright magistrate, for eighteen years one of the selectmen of the town, twenty-seven years town clerk and treasurer, fifteen years a member of the Colonial and State Legislature, and a prominent, honored, and useful member and officer of the church."

In January of the year 1820, Dr. Morse, realizing

that it would be for the best interests of all concerned to relinquish his pastorate at Charlestown, turned his active brain in another direction, and resolved to carry out a plan which he had long contemplated. This was to secure from the Government at Washington an appointment as commissioner to the Indians on the borders of the United States of those early days, in order to enquire into their condition with a view to their moral and physical betterment. To this end he journeyed to Washington and laid his project before the President and the Secretary of War, John C. Calhoun. He was most courteously entertained by these gentlemen and received the appointment.

In the following spring with his son Richard he travelled through the northwestern frontiers of the United States, and gained much valuable information which he laid before the Government. As he was a man of delicate constitution, we cannot but admire his indomitable spirit in ever devising new projects of usefulness to his fellow men. It was impossible for him to remain idle.

But it is not within the scope of this work to follow him on his journeys, although his letters of that period make interesting reading. While he was in Washington his wife, writing to him on January 27, 1820, says: "Mrs. Salisbury and Abby drank tea with us day before yesterday. They told us that Catherine Breese was married to a lieutenant in the army. This must have been a very sudden thing, and I should suppose very grievous to Arthur."

Little did the good lady think as she penned these words that, many years afterwards, her beloved eldest son would take as his second wife a daughter of this union. Why this marriage should have been "grievous" to the father, Arthur Breese, I do not know, unless all army officers were classed among the ungodly by the very pious of those days. As a matter of fact, Lieutenant, afterwards Captain, Griswold was a most gallant gentleman.

In the mean time Finley Morse had reached Charleston in safety after a tedious journey of many days by stage from Washington, and was busily employed in painting. On February 4, 1820, he writes to his mother: —

"I received your good letter of the 19th and 22d ult., and thank you for it. I wish I had time to give you a narrative of my journey as you wish, but you know '*time is money,*' and we must '*make hay while the sun shines,*' and '*a penny saved is a penny got,*' and '*least said soonest mended,*' and a good many other wise sayings which would be quite pat, but I can't think of them.

"The fact is I have scarcely time to say or write a word. I am busily employed in getting the cash, or else Ned's almanac for March will foretell falsely.

"I am doing well, although the city fairly swarms with painters. I am the only one that has as much as he can do; all the rest are complaining. I wish I could divide with some of them, very clever men who have families to support, and can get nothing to do. . . . I feel rejoiced that things have come to such a crisis in Charlestown that our family will be released from that region of trouble so soon.

"Keep up your spirits, mother, the Lord will show you good days according to those in which you have seen evil. . . .

"I am glad Lucretia and the dear little Susan intend

meeting me at New Haven. I think this by far the best plan; it will save me a great deal of time, which, as I said before, is money.

"I shall have to spend some time in New Haven getting settled, and I wish to commence painting as soon as possible, for I have more than a summer's work before me in the President's portrait and Mrs. Ball's.

"As soon as the cash comes in, mother, it shall all be remitted except what I immediately want. You may depend upon it that nothing shall be left undone on my part to help you and the rest of us from that hole of vipers.

"I think it very probable I shall return by the middle of May; it will depend much on circumstances, however. I wish very much to be with my dear wife and daughter. I must contrive to bring them with me next season to Charleston, though it may be more expensive, yet I do not think that should be a consideration. I think that a man should be separated from his family but very seldom, and then under cases of absolute necessity, as I consider the case to be at present with me: that is, I think they should not be separated for any length of time. If I know my own disposition I am of a domestic habit, formed to this habit, probably, by the circumstances that have been so peculiar to our family in Charlestown. I by no means regret having such a habit if it can be properly regulated. I think it may be carried to excess, and shut us from the opportunities of doing good by mixing with our fellow men."

This pronouncement was very characteristic of the man. He was always, all through his long life, happiest when at home surrounded by all his family, and yet he never shirked the duty of absenting himself from home, even for a prolonged period, when by so doing he could accomplish some great or good work.

That a portrait-painter's lot is not always a happy one may be illustrated by the following extracts from letters of Morse to the Mrs. Ball whom he mentions in the foregoing letter to his mother, and who seems to have been a most capricious person, insisting on continual alterations, and one day pleased and the next almost insulting in her censure: —

MADAM, — Supposing that I was dealing not only with a woman of honor, but, from her professions, with a Christian, I ventured in my note of the 18th inst., to make an appeal to your conscience in support of the justness of my demand of the four hundred dollars still due from you for your portrait. By your last note I find you are disposed to take an advantage of that circumstance of which I did not suppose you capable. My sense of the justness of my demand was so strong, as will appear from the whole tenor of that note, that I venture this appeal, not imagining that any person of honor, of the least spark of generous feeling, and more especially of Christian principle, could understand anything more than the enforcing my claim by an appeal to that principle which I knew should be the strongest in a real Christian.

Whilst, however, you have chosen to put a different construction on this part of the note, and supposed that I left you to say whether you would pay me anything or nothing, you have (doubtless unconsciously) shown that your conscience has decided in favor of the whole amount which is my due, and which I can never voluntarily relinquish.

You affirm in the first part of your note that, after due consideration, you think the real value of the picture is four hundred dollars (without the frame), yet, had your crop been good, your conscience would have adjudged me the remaining four hundred dollars without hesitation; and again (if your crop should be good) you could pay me the four hundred dollars next season.

Must I understand from this, madam, that the goodness or badness of your crop is the scale on which your conscience measures your obligation to pay a just debt, and that it contracts or expands as your crop increases or diminishes? Pardon me, madam, if I say that this appears to be the case from your letter.

My wish throughout this whole business has been to accommodate the time and terms of payment as much to your convenience as I could consistently with my duty to my family and myself. As a proof of this you need only advert to my note of yesterday, in which I inform you that I am paying interest on money borrowed for the use of my family which your debt, if it had been promptly paid, would have prevented.

And in another letter he says: —

"I completed your picture in the summer with two others which have given, as far as I can learn, entire satisfaction. Yours was painted with the same attention and with the same ability as the others, and admired as a picture, after it was finished, as much by some as the others, and more by many.

"Among these latter were the celebrated Colonel Trumbull and Vanderlyn, painters of New York. . . . You cannot but recollect, madam, that when you yourself with your children visited it, notwithstanding you expressed yourself before them in terms so strong against it and so wounding to my feelings, yet all your children dissented from you, the youngest saying it was 'mama,' and the eldest, 'I am sure, mother, it is very like you.' . . .

"Your picture, from the day I commenced it, has been the source of one of my greatest trials, and, if it has taught me in any degree patience and forbearance, I shall have abundant reason to be thankful for the affliction."

In the end he consented to take less than had been agreed upon in order to close the incident.

As a happy contrast to this episode we have the following quotation from a letter to his wife written on February 17, 1820: —

"Did I tell you in my last that Colonel Alston insisted on giving me *two hundred dollars* more than I asked for the picture of little Sally, and a commission to paint her again full length next season, smaller than the last and larger than the first portrait, for which I shall receive four hundred dollars? He intimates also that I am to paint a picture annually for him. Is not he a strange man? (as people say here). I wish some more of the great fortunes in this part of the country would be as strange and encourage other artists who are men of genius and starving for want of employment."

Morse returned to the North in the spring of 1820 and joined his mother and his wife and daughter in New Haven, where they had preceded him and where they were comfortably and agreeably settled, as will appear from the following sentence in a letter to his good friend and mentor, Henry Bromfield, of London, dated August, 1820: "You will perceive by the heading of this letter that I am in New Haven. My father and his family have left Charlestown, Massachusetts, and are settled in this place. My own family also, consisting of wife and daughter, are pleasantly settled in this delightful spot. I have built me a fine painting-room attached to my house in which I paint my large pictures in the summer, and in the winter I migrate to Charleston, South Carolina, where I have commissions sufficient to employ me for some years to come."

He returned to Charleston in the fall of 1820 and was again compelled to go alone. He writes to his wife on December 27: "I feel the separation this time more than ever, and I felt the other day, when I saw the steamship start for New York, that I had almost a mind to return in her."

From this sentence we learn that the slow schooner of the preceding years had been supplanted by the more rapid steamship, but that is, unfortunately, all he has to say of this great step forward in human progress.

Further on in this same letter he says: "I am occupied fully so that I have no reason to complain. I have not a *press* like the first season or like the last, but still I can say I am all the time employed. . . . My President pleases very much; I have heard no dissatisfaction expressed. It is placed in the great Hall in a fine light and place. . . . Mrs. Ball wants some alterations, that is to say every five minutes she would like it to be different. She is the most unreasonable of all mortals; derangement is her only apology. I can't tell you all in a letter, must wait till I see you. I shall get the rest of the cash from her shortly."

Just at this time the wave of prosperity on which the young man had so long floated, began to subside, for he writes to his wife on January 28, 1821: —

"I wish I could write encouragingly as to my professional pursuits, but I cannot. Notwithstanding the diminished price and the increase of exertion to please, and although I am conscious of painting much better portraits than formerly (which, indeed, stands to reason if I make continual exertion to improve), yet with all I receive no new commissions, cold and procrastinating answers from those to whom I write and who had put their names on my list. I give less satisfaction to those whom I have painted; I receive less attention also from some of those who formerly paid me much attention, and none at all from most."

But with his usual hopefulness he says later on in this letter: —

"Why should I expect my sky to be perpetually unclouded, my sun to be never obscured? I have thus far enjoyed more of the sunshine of prosperity than most of my fellow men. 'Shall I receive good at the hands of the Lord and shall I not also receive evil?'"

In this letter, a very long one, he suggests the establishment of an academy or school of painting in New Haven, so that he may be enabled to live at home with his family, and find time to paint some of the great historical works which he still longed to do. He also tells of the formation of such an academy in Charleston: —

"Since writing this there has been formed here an Academy of Arts to be erected immediately. J. R. Poinsett, Esq., is President, and six others with myself are chosen Directors. What this is going to lead to I don't know. I heard Mr. Cogdell say that it was intended to have lectures read, among other things. I feel not very sanguine as to its success, still I shall do all in my power to help it on as long as I am here."

His forebodings seem to have been justified, for Mr. John S. Cogdell, a sculptor, thus writes of it in later years to Mr. Dunlap: —

"The Legislature granted a charter, but, my good sir, as they possessed no powers under the constitution to confer taste or talent, and possessed none of those feelings which prompt to patronage, they gave none to the infant academy. . . . The institution was allowed from apathy and opposition to die; but Mr. Poinsett and myself with a few others have purchased, with a hope of reviving, the establishment."

Referring to this academy the wife in New Haven, in a letter of February 25, 1821, says: "Mr. Silliman says he is not much pleased to hear that they have an academy for painting in Charleston. He is afraid they will decoy you there."

On March 11, 1821, Morse answers thus: "Tell Mr. Silliman I have stronger *magnets* at New Haven than any academy can have, and, while that is the case, I cannot be decoyed permanently from home."

I wonder if he used the word "magnets" advisedly, for it was with Professor Silliman that he at that time pursued the studies in physics, including electricity, which had so interested him while in college, and it was largely due to the familiarity with the subject which he then acquired that he was, in later years, enabled successfully to perfect his invention.

On the 12th of March, 1821, another daughter was

born to the young couple, and was named Elizabeth Ann after her paternal grandmother. The child lived but a few days, however, much to the grief of her parents and grandparents.

Charleston had now given all she had to give to the young painter, and he packed his belongings to return home with feelings both of joy and of regret. He was overjoyed at the prospect of so soon seeing his dearly loved wife and daughter, and his parents and brothers; at the same time he had met with great hospitality in Charleston; had made many firm friends; had impressed himself strongly on the life of the city, as he always did wherever he went, and had met with most gratifying success in his profession. A partial list of the portraits painted while he was there gives the names of fifty-five persons, and, as the prices received are appended, we learn that he received over four thousand dollars from his patrons for these portraits alone.

On March 31, 1821, he joyfully announces his home-coming: "I just drop you a hasty line to say that, in all probability, your husband will be with you as soon, if not sooner than this letter. I am entirely clear of all sitters, having outstayed my last application; have been engaged in finishing off and packing up for two days past and contemplate embarking by the middle or end of the coming week in the steamship for New York. You must not be surprised, therefore, to see me soon after this reaches you; still don't be disappointed if I am a little longer, as the winds most prevalent at this season are head winds in going to the North. I am busy in collecting my dues and paying my debts."

CHAPTER XII

MAY 23, 1821 — DECEMBER 17, 1824

Accompanies Mr. Silliman to the Berkshires. — Takes his wife and daughter to Concord, New Hampshire. — Writes to his wife from Boston about a bonnet. — Goes to Washington, D.C. — Paints large picture of House of Representatives. — Artistic but not financial success. — Donates five hundred dollars to Yale. — Letter from Mr. DeForest. — New York "Observer." — Discouragements. — First son born. — Invents marble-carving machine. — Goes to Albany. — Stephen Van Rensselaer. — Slight encouragement in Albany. — Longing for a home. — Goes to New York. — Portrait of Chancellor Kent. — Appointed attaché to Legation to Mexico. — High hopes. — Takes affecting leave of his family. — Rough journey to Washington. — Expedition to Mexico indefinitely postponed. — Returns North. — Settles in New York. — Fairly prosperous.

Much as Morse longed for a permanent home, where he could find continuous employment while surrounded by those he loved, it was not until many years afterwards and under totally different circumstances that his dream was realized. For the present the necessity of earning money for the support of his young family and for the assistance of his ageing father and mother drove him continually forth to new fields, and on May 23, 1821, which must have been only a few weeks after his return from the South, he writes to his wife from Pittsfield, Massachusetts: —

"We are thus far on our tour safe and sound. Mr. Silliman's health is very perceptibly better already. Last night we lodged at Litchfield; Mr. Silliman had an excellent night and is in fine spirits.

"At Litchfield I called on Judge Reeves and sat a little while. . . . I called at Mr. Beecher's with Mr. Silliman and Judge Gould; no one at home. Called with Mr.

Silliman at Dr. Shelden's, and stayed a few moments; sat a few moments also at Judge Gould's.

"I was much pleased with the exterior appearance of Litchfield; saw at a distance Edwards's pickerel pond.

"We left at five this morning, breakfasted at Norfolk, dined at Stockbridge. We there left the stage and have hired a wagon to go on to Middlebury, Vermont, at our leisure. We lodge here to-night and shall probably reach Bennington, Vermont, to-morrow night.

"I have made one slight pencil sketch of the Hoosac Mountain. At Stockbridge we visited the marble quarries, and to-morrow at Lanesborough shall visit the quarries of fine white marble there.

"I am much delighted with my excursion thus far. To travel with such a companion as Mr. Silliman I consider as highly advantageous as well as gratifying."

This is all the record I have of this particular trip. The Mr. Beecher referred to was the father of Henry Ward Beecher.

Later in the summer he accompanied his wife and little daughter to Concord, New Hampshire, and left them there with her father and mother. Writing to her from Boston on his way back to New Haven, he says in characteristically masculine fashion: —

"I have talked with Aunt Bartlett about getting you a bonnet. She says that it is no time to get a fashionable winter bonnet in Boston now, and that it would be much better if you could get it in New York, as the Bostonians get their fashions from New York and, of course, much later than we should in New Haven. She thinks that white is better than blue, etc., etc., etc., which she can explain to you much better than I can. She is willing, however, to get you any you wish if you still request it. She thinks, if you cannot wait for the new fashion, that your black bonnet put into proper shape with black plumes would be as *tasty* and fashionable as any you could procure. I think so, too. You had better write Aunt particularly about it."

While Morse had conscientiously tried to put the best of himself into the painting of portraits, and had succeeded better than he himself knew, he still longed for wider fields, and in November, 1821, he went to Washington, D.C., to begin a work which he for some time had had in contemplation, and which he now felt justified in undertaking. This was to be a large painting of the House of Representatives with many portraits of the members. The idea was well received at Washington and he obtained the use of one of the rooms at the Capitol for a studio, making it easy for the members to sit for him. It could not have been all plain sailing, however, for his wife says to him in a letter of December 28, 1821: "Knowing that perseverance is a trait in your character, we do not any of us feel surprised to hear you have overcome so many obstacles. You have undertaken a great work. . . . Every one thinks it must be a very popular subject and that you will make a splendid picture of it."

Writing to his wife he says: —

"I am up at daylight, have my breakfast and prayers over and commence the labors of the day long before the workmen are called to work on the Capitol by the bell. This I continue unremittingly till one o'clock, when I dine in about fifteen minutes and then pursue my labors until tea, which scarcely interrupts me, as I often have my cup of tea in one hand and my pencil in the other. Between ten and eleven o'clock I retire to rest. This has been my course every day (Sundays, of course, excepted) since I have been here, making about fourteen hours' study out of the twenty-four.

"This you will say is too hard, and that I shall injure my health. I can say that I never enjoyed better health, and my body, by the simple fare I live on, is disciplined to this course. As it will not be necessary to continue long so assiduously I shall not fail to pursue it till the work is done.

"I receive every possible facility from all about the Capitol. The doorkeeper, a venerable man, has offered to light the great chandelier expressly for me to take my sketches in the evening for two hours together, for I shall have it a candlelight effect, when the room, already very splendid, will appear ten times more so."

On the 2d of January, 1822, he writes: "I have commenced to-day taking the likenesses of the members. I find them not only willing to sit, but apparently esteeming it an honor. I shall take seventy of them and perhaps more; all if possible. I find the picture is becoming the subject of conversation, and every day gives me greater encouragement. I shall paint it on part of the great canvas when I return home. It will be eleven feet by seven and a half feet. . . . It will take me until October next to complete it."

The room which he painted was then the Hall of Representatives, but is now Statuary Hall. As a work of art the painting is excellent and is highly esteemed by artists of the present day. It contains eighty portraits.

His high expectations of gaining much profit from its

exhibition and of selling it for a large sum were, however, doomed to disappointment. It did not attract the public attention which he had anticipated and it proved a financial loss to him. It was finally sold to an Englishman, who took it across the ocean, and it was lost sight of until, after twenty-five years, it was found by an artist friend, Mr. F. W. Edmonds, in New York, where it had been sent from London. It was in a more or less damaged condition, but was restored by Morse. It eventually became the property of the late Daniel Huntington, who loaned it to the Corcoran Gallery of Art in Washington, where it now hangs.[1]

I find no more letters of special interest of the year 1822, but Mr. Prime has this to record: "In the winter of 1822, notwithstanding the great expenses to which Mr. Morse had been subjected in producing this picture, and before he had realized anything from its exhibition, he made a donation of five hundred dollars to the library fund of Yale College; probably the largest donation in proportion to the means of the giver which that institution ever received."

The corporation, by vote, presented the thanks of the board in the following letter: —

YALE COLLEGE,
December 4th, 1822.

DEAR SIR, — I am directed by the corporation of this college to present to you the thanks of the board for your subscription of five hundred dollars for the enlargement of the library. Should this example of liberality be generally imitated by the friends of the institution, we

[1] This painting has recently been purchased by the Trustees of the Corcoran Gallery.

should soon have a library creditable to the college and invaluable to men of literary and philosophic research.

With respectful and grateful acknowledgment,

Your obedient servant,

JEREMIAH DAY.

While he was at home in New Haven in the early part of 1823 he sought orders for portraits, and that he was successful in at least one instance is evidenced by the following letter: —

Mr. D. C. DeForest's compliments to Mr. Morse. Mr. DeForest desires to have his portrait taken such as it would have been six or eight years ago, making the necessary calculation for it, and at the same time making it a good likeness in all other respects.

This reason is not to make himself younger, but to appear to children and grandchildren more suitably matched as to age with their mother and grandmother.

If Mr. Morse is at leisure and disposed to undertake this work, he will please prepare his canvas and let me know when he is ready for my attendance.

NEW HAVEN,
30th March, 1823.

Whether Morse succeeded to the satisfaction of Mr. DeForest does not appear from the correspondence, but both this portrait and that of Mrs. DeForest now hang in the galleries of the Yale School of the Fine Arts, and are here reproduced so that the reader may judge for himself.

On the 17th of May, 1823, the first number of the New York "Observer" was published. While being a religious newspaper the prospectus says it "contains also miscellaneous articles and summaries of news and information on every subject in which the community is interested."

This paper was founded and edited by the two brothers Sidney E. and Richard C. Morse, who had abandoned respectively the law and the ministry. It was very successful, and became at one time a power in the community and is still in existence.

The editorial offices were first established at 50 Wall Street, but later the brothers bought a lot and erected a building at the corner of Nassau and Beekman Streets, and that edifice had an important connection with the invention of the telegraph. On the same site now stands the Morse Building, a pioneer sky-scraper now sadly dwarfed by its gigantic neighbors.

The year 1823 was one of mingled discouragement and hope. Compelled to absent himself from home for long periods in search of work, always hoping that in some place he would find enough to do to warrant his bringing his family and making for them a permanent home, his letters reflect his varying moods, but always with the underlying conviction that Providence will yet order all things for the best. The letters of the young wife are pathetic in their expressions of loneliness during the absence of her husband, and yet of forced cheerfulness and submission to the will of God.

On the 17th of March, 1823, another child was born, a son, who was named for his maternal grandfather, Charles Walker. The child was at first very delicate, and this added to the anxieties of the fond mother and father, but he soon outgrew his childish ailments.

MR. D. C. DE FOREST

From "Thistle Prints." Copyright Detroit Publishing Co.

MRS. D. C. DE FOREST

From a painting by Morse now in the Gallery of the Yale School of the Fine Arts

Morse's active mind was ever bent on invention, and in this year he devised and sought to patent a machine for carving marble statues, "perfect copies of any model." He had great hopes of pecuniary profit from this invention and it is mentioned many times in the letters of this and the following year, but he found, on enquiry, that it was not patentable, as it would have been an infringement on the machine of Thomas Blanchard which was patented in 1820.

So once more were his hopes of independence blasted, as they had been in the case of the pump and fire-engine. He longed, like all artists, to be free from the petty cares and humiliations of the struggle for existence, free to give full rein to his lofty aspirations, secure in the confidence that those he loved were well provided for; but, like most other geniuses, he was compelled to drink still deeper of the bitter cup, to drain it to the very dregs.

In the month of August, 1823, he went to Albany, hoping through his acquaintance with the Patroon, Stephen Van Rensselaer, to establish himself there. He painted the portrait of the Patroon, confident that, by its exhibition, he would secure other orders. In a letter to his wife he says: —

"I have found lodgings — a large front room on the second story, twenty-five by eighteen feet, and twelve feet high — a fine room for painting, with a neat little bedroom, and every convenience, and board, all for six dollars a week, which I think is very reasonable. My landlord is an elderly Irish gentleman with three daughters, once in independent circumstances but now reduced. Everything bears the appearance of old-fashioned gentility which you know I always liked. Everything is neat and clean and genteel. . . . Bishop Hobart and a great many acquaintances were on board of the boat upon which I came up to this city.

"I can form no idea as yet of the prospect of success in my profession here. If I get enough to employ me I shall go no farther; if not, I may visit some of the smaller towns in the interior of the State. I await with some anxiety the result of experiments with my machine. I hope the invention may enable me to remain at home."

"*16th of August.* I have not as yet received any application for a portrait. Many tell me I have come at the wrong time — the same tune that has been rung in my ears so long. I hope the right time will come by and by. The winter, it is said, is the proper season, but, as it is better in the South at that season and it will be more profitable to be there, I shall give Albany a thorough trial and do my best. If I should not find enough to employ me here, I think I shall return to New York and settle there. This I had rather not do at present, but it may be the best that I can do. Roaming becomes more and more irksome. Imperious necessity alone drives me to this course. Don't think by this I am faint-hearted; I shall persevere in this course, painful as is the separation from my family, until Providence clearly points out my duty to return."

"*August 22.* I have something to do. I have one portrait in progress and the promise of more. One hundred dollars will pay all my expenses here for three months, so that the two I am now painting will clear me in that respect and all that comes after will be clear gain. I am, therefore, easier in my mind as to this. The portrait I am now painting is Judge Moss Kent, brother of the

Chancellor. He says that I shall paint the Chancellor when he returns to Albany, and his niece also, and from these particulars you may infer that I shall be here for some little time longer, just so long as my good prospects continue; but, should they fail, I am determined to try New York City, and sit down there in my profession permanently. I believe I have now attained sufficient proficiency to venture there. My progress may be slow at first, but I believe it will be sure. I do not like going South and I have given up the idea of New Orleans or any Southern city, at least for the present. Circumstances may vary this determination, but I think a settlement in New York is more feasible now than ever before. I shall be near you and home in cases of emergency, and in the summer and sickly season can visit you at New Haven, while you can do the same to me in New York until we live again at New Haven altogether. I leave out of this calculation the *machine for sculpture*. If that should entirely succeed, my plans would be materially varied, but I speak of my present plan as if that had failed."

"*August 24*. I finished Mr. Kent's picture yesterday and received the money for it. . . . Mr. Kent is very polite to me, and has introduced me to a number of persons and families, among others to the Kanes — very wealthy people — to Governor Yates, etc. Mr. Clinton's son called on me and invited me to their house. . . . I have been introduced to Señor Rocafuerto, the Spaniard who made so excellent a speech before the Bible Society last May. He is a very handsome man, very intelligent, full of wit and vivacity. He is a great favorite with the ladies and is a man of wealth and a zealous patriot, studying our manners, customs, and improvements, with a view of benefiting his own countrymen in Peru. . . . I long to be with you again and to see you all at *home*. I fear I dote on *home* too much, but mine is such an uncommon home, such a delightful home, that I cannot but feel strongly my privation of its pleasures."

"*August 27*. My last two letters have held out to you some encouraging prospects of success here, but now they seem darkened again. I have had nothing to do this week thus far but to wait patiently. I have advertised in both of the city papers that I should remain one week to receive applications, but as yet it has produced no effect. . . .

"Chancellor Kent is out of town and I was told yesterday would not be in until the end of next month. If I should have nothing to do in the mean time it is hardly worth while to stay solely for that. Many have been talking of having their portraits painted, but there it has thus far ended. I feel a little perplexed to know what to do. I find nothing in Albany which can profitably employ my leisure hours. If there were any pictures or statuary where I could sketch and draw, it would be different. . . . I have visited several families who have been very kind to me, for which I am thankful. . . .

"I shall leave Albany and return to New York a week from to-day if there is no change in my prospects. . . . The more I think of making a push at New York as a permanent place of residence in my profession, the more proper it seems that it should be pretty soon. There is now no rival that I should fear; a few more years may produce one that would be hard to overcome. New York does not yet feel the influx of wealth from the

Western canal but in a year or two she will feel it, and it will be advantageous to me to be previously identified among her citizens as a painter.

"It requires some little time to become known in such a city as New York. Colonel T—— is growing old, too, and there is no artist of education sufficiently prominent to take his place as President of the Academy of Arts. By becoming more known to the New York public, and exerting my talents to discover the best methods of promoting the arts and writing about them, I may possibly be promoted to his place, where I could have a better opportunity of doing *something for the arts in our country*, the object at which I aim."

"*September 3.* I have nothing to do and shall pack up on the morrow for New York unless appearances change again. I have not had full employment since I have been in Albany and I feel miserable in doing nothing. I shall set out on Friday, and perhaps may go to New Haven for a day or two to look at you all."

He did manage to pay a short visit to his home, and then he started for New York by boat, but was driven by a storm into Black Rock Harbor and continued his journey from there by land. Writing home the day after his arrival he says: "I have obtained a place to board at friend Coolidge's at two dollars and twenty-five cents a week, and have taken for my studio a fine room in Broadway opposite Trinity Churchyard, for which I am to pay six dollars and fifty cents a week, being fifty cents less than I expected to pay."

There has been some increase in the rental price of rooms on Broadway opposite Trinity Churchyard since that day.

Further on he says: —

"I shall go to work in a few days vigorously. It is a half mile from my room to the place where I board, so that I am obliged to walk more than three miles every day. It is good exercise for me and I feel better for it. I sleep in my room on the floor and put my bed out of sight during the day, as at Washington. I feel in the spirit of 'buckling down to it,' and am determined to paint and study with all my might this winter."

The loving wife is distressed at the idea of his sleeping on the floor, and thus expresses herself in a letter which is dated, curiously enough, November 31: "You know, dear Finley, I have always set my face as a flint and have borne my testimony against your sleeping on the floor. Indeed, it makes my heart ache, when I go to bed in my comfortable chamber, to think of my dear husband sleeping without a bedstead. Your mother says she sent one to Richard, which he has since told her was unnecessary as he used a settee, and which you can get of him. But, if it is in use, do get one or I shall take no comfort."

Soon after his arrival in New York he began the portrait of Chancellor Kent, and writing of him he says: —

"He is not a good sitter; he scarcely presents the same view twice; he is very impatient and you well know that I cannot paint an impatient person; I must have my mind at ease or I cannot paint.

"I have no more applications as yet, but it is not time to expect them. All the artists are complaining, and there are many of them, and they are all poor. The arts are as low as they can be. It is no better at the South, and all the accounts of the arts or artists are of the most discouraging nature."

The portrait of the Chancellor seems not to have brought him more orders, for a little later he writes to his wife: "I waited many days in the hope of some application in my profession, but have been disappointed until last evening I called and spent the evening with my friend Mr. Van Schaick, and told him I had thought of painting some little design from the 'Sketch Book,' so as not to be idle, and mentioned the subject of Ichabod Crane discovering the headless horseman.

"He said: 'Paint it for me and another picture of the same size, and I will take them of you.' So I am now employed. . . .

"*My secret scheme* is not yet disclosable, but I shall let you know as soon as I hear anything definite."

Still later he says: —

"I have seen many of the artists; they all agree that little is doing in the city of New York. It seems wholly given to commerce. Every man is driving at one object — the making of money — not the spending of it. . . .

"My *secret scheme* looks promising, but I am still in suspense; you shall know the moment it is decided one way or the other."

His brother, Sidney Edwards, in a letter to his parents of December 9, 1823, says: "Finley is in good spirits again; not because he has any prospect of business here, but he is dreaming of the gold mines of Mexico."

As his *secret* was now out, he explains it fully in the following letter to his wife, dated December 21, 1823: —

"My cash is almost gone and I begin to feel some anxiety and perplexity to know what to do. I have advertised, and visited, and hinted, and pleaded, and even asked one man to sit, but all to no purpose. . . . My expenses, with the most rigid economy, too, are necessarily great; my rent to-morrow will amount to thirty-three dollars, and I have nothing to pay it with.

"What can I do? I have been here five weeks and there is not the smallest prospect *now* of any difference as to business. I am willing to stay and wish to stay if there is anything to do. The pictures that I am painting for Mr. Van Schaick will not pay my expenses if painted here; my rent and board would eat it all up.

"I have thought of various plans, but what to decide upon I am completely at a loss, nor can I decide until I hear definitely from Washington in regard to my Mexico expedition. Since Brother Sidney has hinted it to you I will tell you the state of it. I wrote to General Van Rensselaer, Mr. Poinsett, and Colonel Hayne, of the Senate, applying for some situation in the legation to Mexico soon to be sent thither. I stated my object in going and my wish to go free of expense and under government protection.

"I received a letter a few days ago from General Van Rennselaer in which he says: 'I immediately laid your request before the President and seconded it with my warmest recommendations. It is impossible to predict the result at present. If our friend Mr. Poinsett is appointed minister, which his friends are pressing, he will no doubt be happy to have you in his suite.'

"Thus the case rests at present. If Mr. Poinsett is appointed I shall probably go to Mexico, if not, it will be more doubtful. . . . If I go I should take my picture of the House of Representatives, which, in the present state of favorable feeling towards our country, I should probably dispose of to advantage.

"All accounts that I hear from Mexico are in the highest degree favorable to my enterprise, and I hear much from various quarters."

As can well be imagined, his wife did not look with unalloyed pleasure on this plan. She says in a letter of December 25, 1823: "I have felt much for you, my dearest Finley, in all your trials and perplexities. I was sorry to hear you had been unsuccessful in obtaining portraits. I hope you will, ere long, experience a change for the better. . . . As to the Mexico plan, I know not what to think of it. How can I consent to have you be at such a distance?"

However, convinced by her husband that it would be for his best interests to go, she reluctantly gave her consent and he used every legitimate effort to secure the appointment. He was finally successful. Mr. Poinsett was not appointed as minister; this honor was bestowed on the Honorable Ninian Edwards, of Illinois, but Morse was named as one of his suite.

In a note from the Honorable Robert Young Hayne, who, it will be remembered, was the opponent of Daniel Webster in the great debates on States' Rights in the Senate, Morse was thus apprised of his appointment: "Governor Edwards's suite consists of Mr. Mason, of Georgetown, D.C., secretary of the legation; Mr. Hodgson, of Virginia, private secretary; and yourself, attaché."

Morse had great hopes of increasing his reputation as a painter and of earning much money in Mexico. He was perfectly frank in stating that his principal object in seeking an appointment as attaché was that he might pursue his profession, and, in a letter to Mr. Edwards of April 15, 1824, he thus explains why he considers this not incompatible with his duties as attaché: "That the pursuit of my profession will not be derogatory to the situation I may hold I infer from the fact that many of the ancient painters were ambassadors to different European courts, and pursued their professions constantly while abroad. Rubens, while ambassador to the English court, executed some of his finest portraits and decorated the ceiling of the chapel of White Hall with some of his best historical productions."

When it was finally decided that he should go, he made all his preparations, including a bed and bedding among his impedimenta, being assured that this was necessary in Mexico, and bade farewell to his family.

His father, his wife and children, and his sister-in-law accompanied him as far as New York. Writing of the parting he says: "A thousand affecting incidents of separation from my beloved family crowded upon my recollection. The unconscious gayety of my dear children as they frolicked in all their wonted playfulness, too young to sympathize in the pangs that agitated their distressed parents; their artless request to bring home some trifling toy; the parting kiss, not understood as meaning more than usual; the tears and sad farewells of father, mother, wife, sister, family, friends; the desolateness of every room as the parting glance is thrown on each familiar object, and 'farewell, farewell' seemed written on the very walls,—all these things bear upon my memory, and I realize the declaration that 'the places which now know us shall know us no more.'"

It must be borne in mind that a journey in those days, even one from New York to Washington, was not a few

LUCRETIA PICKERING WALKER, WIFE OF S. F. B. MORSE, AND TWO CHILDREN

Painted by Morse

hours' ride in a luxurious Pullman, but was fraught with many discomforts, delays, and even dangers.

As an example of this I shall quote the first part of a letter written by Morse from Washington to his wife on April 11, 1824: —

"I lose not a moment in informing you of my safe arrival, with all my baggage, in good order last evening. I was much fatigued, went to bed early, and this morning feel perfectly refreshed and much better for my journey.

"After leaving you on Wednesday morning I had but just time to reach the boat before she started. In the land carriage we occupied three stages over a very rough road. In crossing a small creek in a ferry-boat the stage ahead of ours left the boat a little too soon and came near upsetting in the water, which would have put the passengers into a dangerous situation. As it was the water came into the carriage and wet some of the baggage. It was about an hour before they could get the stage out of the water.

"Next came our turn. After travelling a few miles the springs on one side gave way and let us down, almost upsetting us. We got out without difficulty and, in a few minutes, by putting a rail under one side, we proceeded on again, jocosely telling the passengers in the third stage that it was their turn next.

"When we arrived at the boat in the Delaware to our surprise the third stage came in with a rail under one side, having met with a similar accident a few miles after we left them. So we all had our turn, but no injury to any of us."

His high hopes of success in this enterprise were soon

doomed to be shattered, and once again he was made to suffer a bitter disappointment.

On April 19 he writes: "I am at this moment put into a very embarrassing state of suspense by a political occurrence which has caused a great excitement here, and will cause considerable interest, no doubt, throughout the country. This morning a remonstrance was read in the House of Representatives from the Honorable Ninian Edwards against Mr. Crawford, which contains such charges and of so serious a nature as has led to the appointment of a select committee, with power to send for *persons* and papers in order to a full investigation; and I am told by many members of Congress that Mr. Edwards will undoubtedly be sent for, which will occasion, of course, a great delay in his journey to Mexico, if not cause a suspension of his going until the next season."

The Mr. Crawford alluded to was William Harris Crawford, at that time a prominent candidate for the Presidency in the coming election.

With his customary faith in an overruling Providence, Morse says later in the same letter: "This delay and suspense tries me more than distance or even absence from my dear family. If I could be on my way and pursuing my profession I should feel much better. But all will be for the best; though things look dark I can and will trust Him who will make my path of duty plain before me. This satisfies my mind and does not allow a single desponding thought."

The sending of the legation was indefinitely postponed, and Morse, much disappointed but resolved not to be overwhelmed by this crushing of his high hopes, returned to New Haven.

He spent the summer partly at home and partly in Concord, New Hampshire (where his wife and children had gone to visit her father), and in Portsmouth, Portland, and Hartford, having been summoned to those cities by patrons who wished him to paint their portraits.

We can imagine that the young wife did not grieve over the failure of the Mexican trip. Her letters to her husband at that period are filled with expressions of the deepest affection, but with an undertone of melancholy, due, no doubt, to the increasing delicacy of her health, never very robust.

In the fall of 1824 Morse resolved to make another assault on the purses of the solid men of New York, and he established himself at 96 Broadway, where, for a time, he had the satisfaction of having his wife and children with him. They, however, returned later to New Haven, and on December 5, 1824, he writes to his wife: —

"I am fully employed and in excellent spirits. I am engaged in painting the full-length portrait of Mr. Hone's little daughter, a pretty little girl just as old as Susan. I have made a sketch of the composition with which I am pleased, and so are the father and mother. I shall paint her with a cat set up in her lap like a baby, with a towel under its chin and a cap on its head, and she employed in feeding it with a spoon. . . .

"I am as happy and contented as I can be without my dear Lucrece and our dear children, but I hope it will not be long before we shall be able to live together without these separations."

"*December 17, 1824.* I have everything very comfortable at my rooms. My two pupils, Mr. Agate and Mr.

Field, are very tractable and very useful. I have everything 'in Pimlico,' as mother would say.

"I have begun, and thus far carried on, a system of neatness in my painting-room which I never could have with Henry. Everything has its place, and every morning the room is swept and all things put in order. . . .

"I have as much as I can do in painting. I do not mean by this that I have the overflow that I had in Charleston, nor do I wish it. A hard shower is soon over; I wish rather the gentle, steady, continuing rain. I feel that I have a character to obtain and maintain, and therefore my pictures must be carefully studied. I shall not by this method paint so fast nor acquire property so fast, but I shall do what is better, secure a continuance of patronage and success.

"I have no disposition to be a nine days' wonder, all the rage for a moment and then forgotten forever; compelled on this very account to wander from city to city, to shine a moment in one and then pass on to another."

In a letter of a later date he says: —

"I am going on prosperously through the kindness of Providence in raising up many friends who are exerting themselves in my favor. My storms are partly over, and a clear and pleasant day is dawning upon me."

CHAPTER XIII

JANUARY 4, 1825 — NOVEMBER 18, 1825

Success in New York. — Chosen to paint portrait of Lafayette. — Hope of a permanent home with his family. — Meets Lafayette in Washington. — Mutually attracted. — Attends President's levee. — Begins portrait of Lafayette. — Death of his wife. — Crushed by the news. — His attachment to her. — Epitaph composed by Benjamin Silliman. — Bravely takes up his work again. — Finishes portrait of Lafayette. — Describes it in letter of a later date. — Sonnet on death of Lafayette's dog. — Rents a house in Canal Street, New York. — One of the founders of National Academy of Design. — Tactful resolutions on organization. — First thirty members. — Morse elected first president. — Reëlected every year until 1845. — Again made president in 1861. — Lectures on Art. — Popularity.

It is a commonly accepted belief that a particularly fine, clear day is apt to be followed by a storm. Meteorologists can probably give satisfactory scientific reasons for this phenomenon, but, be that as it may, how often do we find a parallel in human affairs. A period of prosperity and happiness in the life of a man or of a nation is almost invariably followed by calamities, small or great; but, fortunately for individuals and for nations, the converse is also true. The creeping pendulum of fate, pausing for an instant at its highest point, dips down again to gather impetus for a higher swing.

And so it was with Morse. Fate was preparing for him a heavy blow, one of the tragedies of his eventful life, and, in order to hearten him for the trial, to give him strength to bear up under it, she cheered his professional path with the sun of prosperity.

Writing to his wife from New York on January 4, 1825, he says: —

"You will rejoice with me, I know, in my continued and increasing success. I have just learned in confidence, from one of the members of the committee of the corporation appointed to procure a full-length portrait of Lafayette, that they have designated me as the painter of it, and that a subcommittee was appointed to wait on me with the information. They will probably call to-morrow, but, until it is thus officially announced to me, I wish the thing kept secret, except to the family, until I write you more definitely on the subject, which I will do the moment the terms, etc., are settled with the committee.

"I shall probably be under the necessity of going to Washington to take it immediately (the corporation, of course, paying my expenses). But of this in my next."

"*January 6, 1825.* I have been officially notified of my appointment to paint the full-length portrait of Lafayette for the City of New York, so that you may make it as public as you please.

"The terms are not definitely settled; the committee is disposed to be very liberal. I shall have at least seven hundred dollars — probably one thousand. I have to wait until an answer can be received from Washington, from Lafayette to know when he can see me. The answer will arrive probably on Wednesday morning; after that I can determine what to do about going on.

"The only thing I fear is that it is going to deprive me of my dear Lucretia. Recollect the old lady's saying, often quoted by mother, 'There is never a convenience but there ain't one'; I long to see you."

It was well for the young man that he did not realize how dreadfully his jesting fears were to be realized.

Further on he says: "I have made an arrangement with Mr. Durand to have an engraving of Lafayette's portrait. I receive half the profits. Vanderlyn, Sully, Peale, Jarvis, Waldo, Inman, Ingham, and some others were my competitors in the application for this picture."

"*January 8.* Your letter of the 5th I have just received, and one from the committee of medical students engaging me to paint Dr. Smith's portrait for them when I come to New Haven. They are to give me one hundred dollars. I have written them that I should be in New Haven by the 1st of February, or, at farthest, by the 6th; so that it is only prolonging for a little longer, my dear wife, the happy meeting which I anticipated for the 25th of this month. Events are not under our own control.

"When I consider how wonderfully things are working for the promotion of the great and *long-desired* event — that of being constantly with my dear family — all unpleasant feelings are absorbed in this joyful anticipation, and I look forward to the spring of the year with delightful prospects of seeing my dear family permanently settled with me in our own hired house here. There are more encouraging prospects than I can trust to paper at present which must be left for your private ear, and which in magnitude are far more valuable than any encouragement yet made known to me. Let us look with thankful hearts to the Giver of all these blessings."

"*Washington, February 8, 1825.* I arrived safely in this city last evening. I find I have no time to lose, as the Marquis will leave here the 23d. I have seen him and am to breakfast with him to-morrow, and to commence his portrait. If he allows me time sufficient I have no fear as to the result. He has a noble face. In this I am disappointed, for I had heard that his features were not good. On the contrary, if there is any truth in expression of character, there never was a more perfect example of accordance between the face and the character. He has all that noble firmness and consistency, for which he has been so distinguished, strongly indicated in his whole face.

"While he was reading my letters I could not but call to mind the leading events of his truly eventful life. 'This is the man now before me, the very man,' thought I, 'who suffered in the dungeon of Olmütz; the very man who took the oaths of the new constitution for so many millions, while the eyes of thousands were fixed upon him (and which is so admirably described in the Life which I read to you just before I left home); the very man who spent his youth, and his fortune, and his time, to bring about (under Providence) our happy Revolution; the friend and companion of Washington, the terror of tyrants, the firm and consistent supporter of liberty, the man whose beloved name has rung from one end of this continent to the other, whom all flock to see, whom all delight to honor; this is the man, the very identical man!' My feelings were almost too powerful for me as I shook him by the hand and received the greeting of — 'Sir, I am exceedingly happy in your acquaintance, and especially on such an occasion.'"

Thus began an acquaintance which ripened into warm friendship between Morse and Lafayette, and which remained unbroken until the death of the latter.

"*February 10, 1825.* I went last night to the President's levee, the last which Mr. Monroe will hold as President of the United States. There was a great crowd and a great number of distinguished characters, among whom were General Lafayette; the President-elect, J. Q. Adams; Mr. Calhoun, the Vice-President elect; General Jackson, etc. I paid my respects to Mr. Adams and congratulated him on his election. He seemed in some degree to shake off his habitual reserve, and, although he endeavored to suppress his feelings of gratification at his success, it was not difficult to perceive that he felt in high spirits on the occasion. General Jackson went up to him and, shaking him by the hand, congratulated him cordially on his election. The General bears his defeat like a man, and has shown, I think, by this act a nobleness of mind which will command the respect of those who have been most opposed to him.

"The excitement (if it may be called such) on this great question in Washington is over, and everything is moving on in its accustomed channel again. All seem to speak in the highest terms of the order and decorum preserved through the whole of this imposing ceremony, and the good feeling which seems to prevail, with but trivial exceptions, is thought to augur well in behalf of the new administration."

(There was no choice by the people in the election of that year, and John Quincy Adams had been chosen President by a vote of the House of Representatives.)

"I went last night in a carriage with four others — Captain Chauncey of the navy; Mr. Cooper, the celebrated author of the popular American novels; Mr. Causici (pronounced Cau-see-chee), the sculptor; and Mr. Owen, of Lanark, the celebrated philanthropist.

"Mr. Cooper remarked that we had on board a more singularly selected company, he believed, than any carriage at the door of the President, namely, a *misanthropist* (such he called Captain Chauncey, brother of the Commodore), a *philanthropist* (Mr. Owen), a *painter* (myself), a *sculptor* (Mr. Causici), and an *author* (himself).

"The Mr. Owen mentioned above is the very man I sometimes met at Mr. Wilberforce's in London, and who was present at the interesting scene I have often related that occurred at Mr. Wilberforce's. He recollected the circumstance and recognized me, as I did him, instantly, although it is twelve years ago.

"I am making progress with the General, but am much perplexed for want of time; I mean *his time*. He is so harassed by visitors and has so many letters to write that I find it exceedingly difficult to do the subject justice. I give him the last sitting in Washington to-morrow, reserving another sitting or two when he visits New York in July next. I have gone on thus far to my satisfaction and do not doubt but I shall succeed entirely, if I am allowed the requisite number of sittings. The General is very agreeable. He introduced me to his son by saying: 'This is Mr. Morse, the painter, the son of the geographer; he has come to Washington to take the topography of my face.' He thinks of visiting New Haven again when he returns from Boston. He regretted not having seen more of it when he was there, as he was much pleased with the place. He remembers Professor Silliman and others with great affection.

"I have left but little room in this letter to express my affection for my dearly loved wife and children; but of that I need not assure them. I long to hear from you, but direct your letters next to New York, as I shall probably be there by the end of next week, or the beginning of the succeeding one.

"Love to all the family and friends and neighbors. Your affectionate husband, as ever."

Alas! that there should have been no telegraph then to warn the loving husband of the blow which Fate had dealt him.

As he was light-heartedly attending the festivities at the White House, and as he was penning these two interesting letters to his wife, letters which she never read, and anticipating with keenest pleasure a speedy reunion, she lay dead at their home in New Haven.

His father thus conveys to him the melancholy intelligence: —

"*February 8th, 1825.* My affectionately beloved Son, — Mysterious are the ways of Providence. My heart is in pain and deeply sorrowful while I announce to you the sudden and unexpected death of your dear and deservedly loved wife. Her disease proved to be an *affection of the heart* — incurable, had it been known. Dr. Smith's letter, accompanying this, will explain all you will desire to know on this subject.

"I wrote you yesterday that she was convalescent. So she then appeared and so the doctor pronounced. She was up about five o'clock yesterday P.M. to have her bed made as usual; was unusually cheerful and social; spoke of the pleasure of being with her dear husband in New York ere long; stepped into bed herself, fell back with a momentary struggle on her pillow, her eyes were immediately fixed, the paleness of death overspread her countenance, and in five minutes more, without the slightest motion, her mortal life terminated.

"It happened that just at this moment I was entering her chamber door with Charles in my arms, to pay her my usual visit and to pray with her. The nurse met me affrighted, calling for help. Your mother, the family, our neighbors, full of the tenderest sympathy and kindness, and the doctors thronged the house in a few minutes. Everything was done that could be done to save her life, but her 'appointed time' had come, and no earthly power or skill could stay the hand of death.

"It was the Lord who gave her to you, the chiefest of all your earthly blessings, and it is He that has taken her away, and may you be enabled, my son, from the heart to say: 'Blessed be the name of the Lord.' . . . The shock to the whole family is far beyond, in point of severity, that of any we have ever before felt, but we are becoming composed, we hope on grounds which will prove solid and lasting.

"I expect this will reach you on Saturday, the day after the one we have appointed for the funeral, when you will have been in Washington a week and I hope will have made such progress in your business as that you will soon be able to return. . . .

"You need not hurry home. Nothing here requires it. We are all well and everything will be taken good care of. Give yourself no concern on that account. Finish your business as well as you will be able to do it after receiving this sad news."

This blow was an overwhelming one. He could not, of course, compose himself sufficiently to continue his work on the portrait of Lafayette, and, having apprised the General of the reason for this, he received from him the following sympathetic letter:—

I have feared to intrude upon you, my dear sir, but want to tell you how deeply I sympathize in your grief — a grief of which nobody can better than me appreciate the cruel feelings.

You will hear from me, as soon as I find myself again near you, to finish the work you have so well begun.

Accept my affectionate and mournful sentiment.

LAFAYETTE.

The day after he received his father's letter he left Washington and wrote from Baltimore, where he stopped over Sunday with a friend, on February 13: —

MY DEAR FATHER, — The heart-rending tidings which you communicated reached me in Washington on Friday evening. I left yesterday morning, spend this day here at Mr. Cushing's, and set out on my return home to-morrow. I shall reach Philadelphia on Monday night, New York on Tuesday night, and New Haven on Wednesday night.

Oh! is it possible, is it possible? Shall I never see my dear wife again?

But I cannot trust myself to write on this subject. I need your prayers and those of Christian friends to God for support. I fear I shall sink under it.

Oh! take good care of her dear children.

Your agonized son,

FINLEY.

Another son had been born to him on January 20, 1825, and he was now left with three motherless children to provide for, and without the sustaining hope of a speedy and permanent reunion with them and with his beloved wife.

Writing to a friend more than a month after the death of his wife, he says: —

"Though late in performing the promise I made you of writing you when I arrived home, I hope you will attribute it to anything but forgetfulness of that promise. The confusion and derangement consequent on such an afflicting bereavement as I have suffered have rendered it necessary for me to devote the first moments of composure to looking about me, and to collecting and arranging the fragments of the ruin which has spread such desolation over all my earthly prospects.

"Oh! what a blow! I dare not yet give myself up to the full survey of its desolating effects. Every day brings to my mind a thousand new and fond connections with dear Lucretia, all now ruptured. I feel a dreadful void, a heart-sickness, which time does not seem to heal but rather to aggravate.

"You know the intensity of the attachment which existed between dear Lucretia and me, never for a moment interrupted by the smallest cloud; an attachment founded, I trust, in the purest love, and daily strengthening by all the motives which the ties of nature and, more especially, of religion, furnish.

"I found in dear Lucretia everything I could wish. Such ardor of affection, so uniform, so unaffected, I never saw nor read of but in her. My fear with regard to the measure of my affection toward her was not that I might fail of 'loving her as my own flesh,' but that I should put her in the place of Him who has said, 'Thou shalt have no other Gods but me.' I felt this to be my

STUDY FOR PORTRAIT OF LAFAYETTE

Now in New York Public Library

greatest danger, and to be saved from this *idolatry* was often the subject of my earnest prayers.

"If I had desired anything in my dear Lucretia different from what she was, it would have been that she had been *less lovely*. My whole soul seemed wrapped up in her; with her was connected all that I expected of happiness on earth. Is it strange, then, that I now feel this void, this desolateness, this loneliness, this heart-sickness; that I should feel as if my very heart itself had been torn from me?

"To any one but those who knew dear Lucretia what I have said might seem to be but the extravagance of an excited imagination; but to you, who knew the dear object I lament, all that I have said must but feebly shadow her to your memory."

It was well for him that he found constant occupation for his hand and brain at this critical period of his life. The Fates had dealt him this cruel blow for some good reason best known to themselves. He was being prepared for a great mission, and it was meet that his soul, like gold, should be purified by fire; but, at the same time, that the blow might not utterly overwhelm him, success in his chosen profession seemed again to be within his grasp.

Writing to his parents from New York, on April 8, 1825, he says: —

"I have as much as I can do, but after being fatigued at night and having my thoughts turned to my irreparable loss, I am ready almost to give up. The thought of seeing my dear Lucretia, and returning home to her, served always to give me fresh courage and spirits whenever I felt worn down by the labors of the day, and now I hardly know what to substitute in her place.

"To my friends here I know I seem to be cheerful and happy, but a cheerful countenance with me covers an aching heart, and often have I feigned a more than ordinary cheerfulness to hide a more than ordinary anguish.

"I am blessed with prosperity in my profession. I have just received another commission from the corporation of the city to paint a common-sized portrait of Rev. Mr. Stanford for them, to be placed in the almshouse."

The loss of his young wife was the great tragedy of Morse's life. Time, with her soothing touch, healed the wound, but the scar remained. Hers must have been, indeed, a lovely character. Professor Benjamin Silliman, Sr., one of her warmest friends, composed the epitaph which still remains inscribed upon her tombstone in the cemetery at New Haven. (See opposite page.)

With a heavy heart, but bravely determining not to be overwhelmed by this crushing blow, Morse took up his work again. He finished the portrait of Lafayette, and it now hangs in the City Hall in New York. Writing of it many years later to a gentleman who had made some enquiries concerning it, he says: —

"In answer to yours of the 8th instant, just received, I can only say it is so long since I have seen the portrait I painted of General Lafayette for the City of New York, that, strange to say, I find it difficult to recall even its general characteristics.

"That portrait has a melancholy interest for me, for it was just as I had commenced the second sitting of the General at Washington that I received the stunning

IN MEMORY OF
LUCRETIA PICKERING
WIFE OF
SAMUEL F. B. MORSE
WHO DIED 7TH OF FEBRUARY A.D. 1825,
AGED 25 YEARS.

SHE COMBINED, IN HER CHARACTER AND PERSON,
A RARE ASSEMBLAGE OF EXCELLENCES:
BEAUTIFUL IN FORM, FEATURES AND EXPRESSION
PECULIARLY BLAND IN HER MANNERS,
HIGHLY CULTIVATED IN MIND,
SHE IRRESISTIBLY DREW ATTENTION, LOVE,
AND RESPECT;
DIGNIFIED WITHOUT HAUGHTINESS,
AMIABLE WITHOUT TAMENESS,
FIRM WITHOUT SEVERITY, AND
CHEERFUL WITHOUT LEVITY,
HER UNIFORM SWEETNESS OF TEMPER
SPREAD PERPETUAL SUNSHINE AROUND
EVERY CIRCLE IN WHICH
SHE MOVED.
"WHEN THE EAR HEARD HER IT BLESSED HER,
WHEN THE EYE SAW HER IT GAVE
WITNESS TO HER."
IN SUFFERINGS THE MOST KEEN,
HER SERENITY OF MIND NEVER FAILED HER;
DEATH TO HER HAD NO TERRORS,
THE GRAVE NO GLOOM.
THOUGH SUDDENLY CALLED FROM EARTH,
ETERNITY WAS NO STRANGER TO HER THOUGHTS,
BUT A WELCOME THEME OF
CONTEMPLATION.
RELIGION WAS THE SUN
THAT ILLUMINED EVERY VIRTUE,
AND UNITED ALL IN ONE
BOW OF BEAUTY.
HERS WAS THE RELIGION OF THE GOSPEL;
JESUS CHRIST HER FOUNDATION,
THE AUTHOR AND FINISHER OF HER FAITH.
IN HIM SHE RESTS, IN SURE
EXPECTATION OF A GLORIOUS
RESURRECTION.

intelligence of Mrs. Morse's death, and was compelled abruptly to suspend the work. I preserve, as a gratifying memorial, the letter of condolence and sympathy sent in to me at the time by the General, and in which he speaks in flattering terms of the promise of the portrait as a likeness.

"I must be frank, however, in my judgment of my own works of that day. This portrait was begun under the sad auspices to which I have alluded, and, up to the close of the work, I had a series of constant interruptions of the same sad character. A picture painted under such circumstances can scarcely be expected to do the artist justice, and as a work of art I cannot praise it. Still, it is a good likeness, was very satisfactory to the General, and he several times alluded to it in my presence in after years (when I was a frequent visitor to him in Paris) in terms of praise.

"It is a full-length, standing figure, the size of life. He is represented as standing at the top of a flight of steps, which he has just ascended upon a terrace, the figure coming against a glowing sunset sky, indicative of the glory of his own evening of life. Upon his right, if I remember, are three pedestals, one of which is vacant as if waiting for his bust, while the two others are surmounted by the busts of Washington and Franklin — the two associated eminent historical characters of his own time. In a vase on the other side is a flower — the helianthus — with its face toward the sun, in allusion to the characteristic stern, uncompromising consistency of Lafayette — a trait of character which I then considered, and still consider, the great prominent trait of that distinguished man."

Morse, like many men who have excelled in one branch of the fine arts, often made excursions into one of the others. I find among his papers many scraps of poetry and some more ambitious efforts, and while they do not, perhaps, entitle him to claim a poet's crown, some of them are worthy of being rescued from oblivion. The following sonnet was sent to Lafayette under the circumstances which Morse himself thus describes: —

"Written on the loss of a faithful dog of Lafayette's on board the steamboat which sank in the Mississippi. The dog, supposing his master still on board, could not be persuaded to leave the cabin, but perished with the vessel.

"Lost, from thy care to know thy master free
  Can we thy self-devotion e'er forget?
'T was kindred feeling in a less degree
  To that which thrilled the soul of Lafayette.
  He freely braved our storms, our dangers met,
Nor left the ship till we had 'scaped the sea.
  Thine was a spark of noble feeling bright
Caught from the fire that warms thy master's heart.
His was of Heaven's kindling, and no small part
  Of that pure fire is his. We hail the light
Where'er it shines, in heaven, in man, in brute;
We hail that sacred light howe'er minute,
  Whether its glimmering in thy bosom rest
  Or blaze full orb'd within thy master's breast."

This was sent to General Lafayette on the 4th of July, 1825, accompanied by the following note: —

"In asking your acceptance of the enclosed poetic trifle, I have not the vanity to suppose it can contribute much to your gratification; but if it shall be considered as an endeavor to show to you some slight return of gratitude for the kind sympathy you evinced towards me at a time of deep affliction, I shall have attained my aim. Gladly would I offer to you any service, but, while a whole nation stands waiting to answer the expression of your smallest wish, my individual desire to serve you can only be considered as contending for a portion of that high honor which all feel in serving you."

Concealing from the world his great sorrow, and bravely striving always to maintain a cheerful countenance, Morse threw himself with energy into his work in New York, endeavoring to keep every minute occupied.

He seems to have had his little daughter with him for a while, for in a letter of March 12, 1825, occurs this sentence: "Little Susan has had the toothache once or twice, and I have promised her a doll if she would have it out to-day — I am this moment stopped by her coming in and showing me the *tooth out*, so I shall give her the doll."

But he soon found that it would be impossible for him to do justice to his work and at same time fulfil his duties as a parent, and for many years afterwards his motherless children found homes with different relatives, but the expense of their keep and education was always borne by their father.

On the 1st of May, 1825, he moved into new quarters, having rented an entire house at No. 20 Canal Street for the sum of four hundred dollars a year, and he says, "My new establishment will be very commodious for my professional studies, and I do not think its being so far '*up town*' will, on the whole, be any disadvantage to me."

"May 26, 1825. I have at length become comfortably

settled and begin to feel at home in my new establishment. All things at present go smoothly. Brother Charles Walker and Mr. Agate join with me in breakfast and tea, and we find it best for convenience, economy, and time to dine from home, — it saves the perplexity of providing marketing and the care of stores, and, besides, we think it will be more economical and the walk will be beneficial."

While success in his profession seemed now assured, and while orders poured in so fast that he gladly assisted some of his less fortunate brother artists by referring his would-be patrons to them, he also took a deep interest in the general artistic movement of the time.

He was, by nature, intensely enthusiastic, and his strong personality ever impressed itself on individuals and communities with which he came in contact. He was a born leader of men, and, like so many other leaders, often so forgetful of self in his eager desire for the general good as to seriously interfere with his material prosperity. This is what happened to him now, for he gave so liberally of himself in the formation of a new artistic body in New York, and in the preparation of lectures, that he encroached seriously on time which might have been more lucratively employed.

His brother Sidney comments on this in a letter to the other brother Richard: "Finley is well and in good spirits, though not advancing very rapidly in his business. He is full of the Academy and of his lectures — can hardly talk on any other subject. I despair of ever seeing him rich or even at ease in his pecuniary circumstances from efforts of his own, though able to do it with so little effort. But he may be in a better way, perhaps, of getting a fortune in his present course than he would be in the laborious path which we are too apt to think is the only road to wealth and ultimate ease."

We have seen that Morse was one of the founders of an academy of art in Charleston, South Carolina, and we have seen that, after his departure from that city, this academy languished and died. Is it an unfair inference that, if he had remained permanently in Charleston, so sad a fate would not have overtaken the infant academy? In support of this inference we shall now see that he was largely instrumental in bringing into being an artistic association, over which he presided for many years, and which has continued to prosper until, at the present day, it is the leading artistic body in this country.

When Morse settled in New York in 1825 there existed an American Academy of Arts, of which Colonel Trumbull, the celebrated painter, was the president. While eminent as a painter, Trumbull seems to have lacked executive ability and to have been rather haughty and overbearing in his manner, for Morse found great dissatisfaction existing among the professional artists and students.

At first it was thought that, by bringing their grievances before the board of directors of the Academy, conditions might be changed, and on the 8th of November, 1825, a meeting was called in the rooms of the Historical Society, and the "New York Drawing Association" was formed, and Morse was chosen to preside over its meetings. It was not intended, at first, that this association should be a rival of the old Academy, but that it should give to its members facilities which were difficult of attainment in the Academy, and should, perhaps, force that institution to become more liberal.

It was not successful in the latter effort, for at a meeting of the Drawing Association on the evening of the 14th of January, 1825, Morse, the president, proposed certain resolutions which he introduced by the following remarks: —

"We have this evening assumed a new attitude in the community; our negotiations with the Academy are at an end; our union with it has been frustrated after every proper effort on our part to accomplish it. The two who were elected as directors from our ticket have signified their non-acceptance of the office. We are therefore left to organize ourselves on a plan that shall meet the wishes of us all.

"A plan of an institution which shall be truly liberal, which shall be mutually beneficial, which shall really encourage our respective arts, cannot be devised in a moment; it ought to be the work of great caution and deliberation and as simple as possible in its machinery. Time will be required for the purpose. We must hear from distant countries to obtain their experience, and it must necessarily be, perhaps, many months before it can be matured.

"In the mean time, however, a preparatory, simple organization can be made, and should be made as soon as possible, to prevent dismemberment, which may be attempted by outdoor influence. On this subject let us all be on our guard; let us point to our public documents to any who ask what we have done and why we have done it, while we go forward minding only our own concerns, leaving the Academy of Fine Arts as much of our thoughts as they will permit us, and, bending our attention to our own affairs, act as if no such institution existed.

"One of our dangers at present is division and anarchy from a want of organization suited to the present exigency. We are now composed of artists in the four arts of design, namely, painting, sculpture, architecture, and engraving. Some of us are professional artists, others amateurs, others students. To the professed and practical artist belongs the management of all things relating to schools, premiums, and lectures, so that amateur and student may be most profited. The amateurs and students are those alone who can contend for the premiums, while the body of professional artists exclusively judge of their rights to premiums and award them.

"How shall we first make the separation has been a question which is a little perplexing. There are none of us who can assume to be the body of artists without giving offence to others, and still every one must perceive that, to organize an academy, there must be the distinction between professional artists, amateurs who are students, and professional students. The first great division should be the body of professional artists from the amateurs and students, constituting the body who are to manage the entire concerns of the institution, who shall be its officers, etc.

"There is a method which strikes me as obviating the difficulty; place it on the broad principle of the formation of any society — universal suffrage. We are now a mixed body; it is necessary for the benefit of all that a separation into classes be made. Who shall make it?

Why, obviously the body itself. Let every member of this association take home with him a list of all the members of it. Let each one select for himself from the whole list *fifteen*, whom he would call professional artists, to be the ticket which he will give in at the next meeting.

"These fifteen thus chosen shall elect not less than *ten*, nor more than *fifteen*, professional artists, in or out of the association, who shall (with the previously elected fifteen) constitute the body to be called the National Academy of the Arts of Design. To these shall be delegated the power to regulate its entire concerns, choose its members, select its students, etc.

"Thus will the germ be formed to grow up into an institution which we trust will be put on such principles as to encourage — not to depress — the arts. When this is done our body will no longer be the Drawing Association, but the National Academy of the Arts of Design, still including all the present association, but in different capacities.

"One word as to the name 'National Academy of the Arts of Design.' Any less name than 'National' would be taking one below the American Academy, and therefore is not desirable. If we were simply the 'Associated Artists,' their name would swallow us up; therefore 'National' seems a proper one as to the arts of design. These are painting, sculpture, architecture, and engraving, while the fine arts include poetry, music, landscape gardening, and the histrionic arts. Our name, therefore, expresses the entire character of our institution and that only."

From this we see that Morse's enthusiasm was tempered with tact and common sense. His proposals were received with unanimous approval, and on the 15th of January, 1826, the following fifteen were chosen: — S. F. B. Morse, Henry Inman, A. B. Durand, John Frazee, William Wall, Charles C. Ingham, William Dunlap, Peter Maverick, Ithiel Town, Thomas S. Cummings, Edward Potter, Charles C. Wright, Mosely J. Danforth, Hugh Reinagle, Gerlando Marsiglia. These fifteen professional artists added by ballot to their number the following fifteen: — Samuel Waldo, William Jewett, John W. Paradise, Frederick S. Agate, Rembrandt Peale, James Coyle, Nathaniel Rogers, J. Parisen, William Main, John Evers, Martin E. Thompson, Thomas Cole, John Vanderlyn (who declined), Alexander Anderson, D. W. Wilson.

Thus was organized the National Academy of Design. Morse was elected its first president and was annually reëlected to that office until the year 1845, when, the telegraph having now become an assured success, he felt that he could not devote the necessary time and thought to the interests of the Academy, and he insisted on retiring.

In the year 1861 he was prevailed upon by Thomas S. Cummings, one of the original academicians, but now a general, to become again the president, and he served in that office for a year. The General, in a letter to Mr. Prime in 1873, says, "and, I may add, was beloved by all."

I shall not attempt to give a detailed account of the early struggles of the Academy, closely interwoven though they be with Morse's life. Those who may be interested in the matter will find them all detailed in

General Cummings' "Records of the National Academy of Design."

Morse prepared and delivered a number of lectures on various subjects pertaining to the fine arts, and most of these have been preserved in pamphlet form. In this connection I shall quote again from the letter of General Cummings before alluded to: —

"Mr. Morse's connection with the Academy was doubtless unfavorable in a pecuniary point of view; his interest in it interfering with professional practice, and the time taken to enable him to prepare his course of lectures materially contributed to favor a distribution of his labors in art to other hands, and it never fully returned to him. His 'Discourse on Academies of Art,' delivered in the chapel of Columbia College, May, 1827, will long stand as a monument of his ability in the line of art literature.

"As an historical painter Mr. Morse, after Allston, was probably the best prepared and most fully educated artist of his day, and should have received the attention of the Government and a share of the distributions in art commissions."

That his efforts were appreciated by his fellow artists and by the cultivated people of New York is thus modestly described in a letter to his parents of November 18, 1825: —

"I mentioned that reputation was flowing in upon me. The younger artists have formed a drawing association at the Academy and elected me their president. We meet in the evenings of three days in a week to draw, and it has been conducted thus far with such success as to have trebled the number of our association and excited the attention and applause of the community. There is a spirit of harmony among the artists, every one says, which never before existed in New York, and which augurs well for the success of the arts.

"The artists are pleased to attribute it to my exertions, and I find in them in consequence expressions and feelings of respect which have been very gratifying to me. Whatever influence I have had, however, in producing this pleasant state of things, I think there was the preparation in the state of mind of the artists themselves. I find a liberal feeling in the younger part of them, and a refinement of manners, which will redeem the character of art from the degradation to which a few dissipated interlopers have, temporarily, reduced it.

"A Literary Society, admission to which must be by unanimous vote, and into which many respectable literary characters of the city have been denied admission, has chosen me a member, together with Mr. Hillhouse and Mr. Bryant, poets. This indicates good feelings towards me, to say the least, and, in the end, will be of advantage, I have no doubt."

CHAPTER XIV

JANUARY 1, 1826 — DECEMBER 5, 1829

Success of his lectures, the first of the kind in the United States. — Difficulties of his position as leader. — Still longing for a home. — Very busy but in good health. — Death of his father. — Estimates of Dr. Morse. — Letters to his mother. — Wishes to go to Europe again. — Delivers address at first anniversary of National Academy of Design. — Professor Dana lectures on electricity. — Morse's study of the subject. — Moves to No. 13 Murray Street. — Too busy to visit his family. — Death of his mother. — A remarkable woman. — Goes to central New York. — A serious accident. — Moral reflections. — Prepares to go to Europe. — Letter of John A. Dix. — Sails for Liverpool. — Rough voyage. — Liverpool.

January 1, 1826

MY DEAR PARENTS, — I wish you all a Happy New Year! Kiss my little ones as a New Year's present from me, which must answer until I visit them, when I shall bring them each a present if I hear good accounts from them. . . .

The new year brings with it many painful reflections to me. When I consider what a difference a year has accomplished in my situation; that one on whom I depended so much for domestic happiness at this time last year gave me the salutations of the season, and now is gone where years are unknown; and when I think how mysteriously I am separated from my little family, and that duty may keep me I know not how much longer in this solitary state, I have much that makes the present season far from being a Happy New Year to me. But, mysterious as things seem in regard to the future, I know that all will be ordered right, and I have a great deal to say of mercy in the midst of judgment, and a thousand unmerited blessings with all my troubles.

But why do I talk of troubles? My cup is overflowing with blessings. As far as outward circumstances are concerned, Providence seems to be opening an honorable and useful course to me. Oh! that I may be able to bear prosperity, if it is his will to bestow it, or be denied it if not accompanied with his blessing. . . .

I am much engaged in my lectures, have completed two, nearly, and hope to get through the four in season for my turn at the Athenæum. These lectures are of great importance to me, for, if well done, they place me alone among the artists; I being the only one who has as yet written a course of lectures in our country. Time bestowed on them is not, therefore, misspent, for they will acquire me reputation which will yield wealth, as mother, I hope, will live to see.

"*January 15, 1826.* On this day I seem to have the only moment in the week in which I can write you, for I am almost overwhelmed by the multitude of cares that crowd upon me. . . . I find that the path of duty, though plain, is not without its roughness. I can say but in one word that the Association of Artists, of whom I am president, after negotiations of some weeks with the Academy of Fine Arts to come into it on terms of mutual benefit, find their efforts unavailing, and have separated and formed a new academy to be called, probably, the National Academy of the Arts of Design. I am at its head, but the cares and responsibility which devolve on me in consequence are more than a balance for the honor. The battle is yet to be fought for the meed of public favor, and were it not that the entire and perfect justness of our cause is clear to me in every point of view, I should retire from a contest which would merely serve to rouse up all the 'old Adam' to no profit; but the cause of the artists seems, under Providence, to be, in some degree, confided to me, and I cannot shrink from the cares and troubles at present put upon me. I have gone forward thus far, asking direction from above, and, in looking around me, I feel that I am in the path of duty. May I be kept in it and be preserved from the temptations, the various and multiplied and complicated temptations, to which I know I shall be exposed. In every step thus far I feel an approving conscience; there is none I could wish to retrace. . . .

"I fear you will think I have but few thoughts for you all at home, and my dear little ones in particular. I do think of them, though, very often, with many a longing to have a home for them under a parent's roof, and all my efforts now are tending distantly to that end; but when I shall ever have a home of my own, or whether it will ever be, I know not. The necessity for a second connection on their account seems pressing, but I cannot find my heart ready for it. I am occasionally rallied on the subject, but the suggestion only reminds me of her I have lost, and a tear is quite as ready to appear as a smile; or, if I can disguise it, I feel a pang within that shows me the wound is not yet healed. It is eleven months since she has gone, but it seems but yesterday."

"*April 18, 1826.* I don't know but you will think I have forgotten how to write letters, and I believe this is the first I have written for six weeks.

"The pressure of my lectures became very great towards the close of them, and I was compelled to bend my whole attention to their completion. I did not expect, when I delivered my first, that I should be able to give more than two, but the importance of going through seemed greater as I advanced, and I was strengthened to accomplish the whole number, and, if I can judge from various indications, I think I have been successful. My audience, consisting of the most fashionable and literary society in the city, regularly increased at each successive lecture, and at the last it was said that I had the largest audience ever assembled in the room.

"I am now engaged on Lafayette in expectation of completing it for our exhibition in May, after which time I hope I shall be able to see you for a day or two in New Haven. I long to see you all, and those dear children often make me feel anxious, and I am often tempted to break away and have a short look at them, but I am tied down here and cannot move at present. All that I am doing has some reference to their interest; they are constantly on my mind.

". . . My health was never better with all my intense application, sitting in my chair from seven in the morning until twelve or one o'clock the next morning, with only about an hour's intermission. I have felt no permanent inconvenience. On Saturday night, generally, I have felt exceedingly nervous, so that my whole body and limbs would shake, but resting on the Sabbath seemed to give me strength for the next week. Since my mind is relieved from my lectures I have felt new life and spirits, and feel strong to accomplish anything."

"*May 10, 1826.* I have just heard from mother and feel anxious about father. Nothing but the most imperious necessity prevents my coming immediately to New Haven; indeed, as it is, I will try and break away sometime next week, if possible, and pass one day with you, but how to do it without detriment to my business I don't know. . . .

"I have longed for some time for a little respite, but, like our good father, all his sons seem destined for most busy stations in society, and constant exertions, not for themselves alone, but for the public benefit."

Whether this promised visit to New Haven was paid or not is not recorded, but it is to be hoped that it was made possible, for the good husband and father, the faithful worker for the betterment of mankind, was called to his well-earned rest on the 9th of June, 1826.

Of him Dr. John Todd said, "Dr. Morse lived before his time and was in advance of his generation." President Dwight of Yale found him "as full of resources as an egg is of meat"; and Daniel Webster spoke of him as "always thinking, always writing, always talking, always acting." Mr. Prime thus sums up his character: "He was a man of genius, not content with what had been and was, but originating and with vast executive ability combining the elements to produce great results. To him more than to any other one man may be attributed the impulses given in his day to religion and learning in the United States. A polished gentleman in his manners; the companion, correspondent, and friend of the most eminent men in Church and State; honored at the early age of thirty-four with the degree of Doctor of Divinity by the University of Edinburgh, Scotland; sought by scholars and statesmen from abroad as one of the foremost men of his country and time."

The son must have felt keenly the loss of his father so soon after the death of his wife. The whole family was a singularly united one, each member depending on the others for counsel and advice, and the father, who was but sixty-five when he died, was still vigorous in mind, although of delicate constitution.

Later in this year Morse managed to spend some time in New Haven, and he persuaded his mother to seek rest and recuperation in travel, accompanying her as far as Boston and writing to her there on his return to New Haven.

"*September 20, 1826.* I arrived safely home after leaving you yesterday and found that neither the house nor the folks had run away. . . . Persevere in your travels, mother, as long as you think it does you good, and tell Dick to brush up his best bows and bring home some lady to grace the now desolate mansion."

On November 9, 1826, he writes to his mother from New York: —

"Don't think I have forgotten you all at home because I have been so remiss in writing you lately. I feel guilty, however, in not stealing some little time just to write you one line. I acknowledge my fault, so please forgive me and I will be a *better boy* in future.

"The fact is I have been engaged for the last three days during all my leisure moments in something unusual with me, — I mean *electioneering*. 'Oh! what a sad boy!' mother will say. 'There he is leaving everything at sixes and sevens, and driving through the streets, and busying himself about those *poison politics.*' Not quite so fast, however.

"I have not neglected my own affairs, as you will learn one of these days. I have an historical picture to paint, which will occupy me for some time, for a proprietor of a steamboat which is building in Philadelphia to be the most splendid ever built. He has engaged historical pictures of Allston, Vanderlyn, Sully, and myself, and landscapes of the principal landscape painters, for a gallery on board the boat. I consider this as a new and noble channel for the encouragement of painting, and in such an enterprise and in such company I shall do my best.

"What do you think of sparing me for about one year to visit Paris and Rome to finish what I began when in Europe before? My education as a painter is incomplete without it, and the time is rapidly going away when my age will render it impossible to profit by such studies, even if I should be able, at a future time, to visit Europe again. . . . I can, perhaps, leave my dear little ones at their age better than if they were more advanced, and, as my views are ultimately to benefit them, I think no one will accuse me of neglecting them. If they do, they know but little of my feelings towards them."

The mother's answer to this letter has not been preserved, but whether she dissuaded him from going at that time, or whether other reasons prevented him, the fact is that he did not start on the voyage to Europe (the return trip proving so momentous to himself and to the world) until exactly three years later.

I shall pass rapidly over these intervening three years. They were years of hard work, but of work rewarded by material success and increasing honor in the community.

On May 3, 1827, on the occasion of the first anniversary of the National Academy of Design, Morse, its president, delivered an address before a brilliant audience in the chapel of Columbia College. This address was considered so remarkable that, at the request of the Academy, it was published in pamphlet form. It called forth a sharp review in the "North American," which voiced the opinions of those who were hostile to the new Academy, and who considered the term "National" little short of arrogant. Morse replied to this attack in a masterly manner in the "Journal of Commerce," and this also was published in pamphlet form and ended the controversy.

In the year 1827, Professor James Freeman Dana, of Columbia College, delivered a series of lectures on the subject of electricity at the New York Athenæum. Professor Dana was an enthusiast in the study of that science, which, at that time, was but in its infancy, and he foresaw great and beneficial results to mankind from this mysterious force when it should become more fully understood.

Morse, already familiar with the subject from his experiments with Professor Silliman in New Haven, took a deep interest in these lectures, and he and Professor Dana became warm friends. The latter, on his side a great admirer of the fine arts, spent many hours in the studio of the artist, discussing with him the two subjects which were of absorbing interest to them both, art and electricity. In this way Morse became perfectly familiar with the latest discoveries in electrical science, so that when, a few years later, his grand conception of a simple and practicable means of harnessing this mystic agent to the uses of mankind took form in his brain, it found a field already prepared to receive it. I wish to lay particular emphasis on this point because, in later years, when his claims as an inventor were bitterly assailed in the courts and in scientific circles, it was asserted that he knew nothing whatever of the science of electricity at the time of his invention, and that all its essential features were suggested to him by others.

In the year 1828, Morse again changed his quarters, moving to a suite of rooms at No. 13 Murray Street, close to Broadway, for which he paid a "great rent," $500, and on May 6 of that year he writes to his mother:

"Ever since I left you at New Haven I have been over head and ears in arrangements of every kind. It is the busiest time of the whole year as it regards the National Academy. We have got through the arrangement of our exhibition and yesterday opened it to the guests of the Academy. We had the first people in the city, ladies and gentlemen, thronging the room all day, and the voice of all seemed to be — 'It is the best exhibition of the kind that has been seen in the city.'

"I am now arranging my rooms; they are very fine ones. I shall be through in a few days, and then I hope to be able to come up and see you, for I feel very anxious about you, my dear mother. I do most sincerely sympathize with you in your troubles and long to come up and take some of the care and burden from you, and will do it as soon as my affairs here can be arranged so that I can leave them without serious detriment to them. . . . What a siege you must have had with your *help*, as it is most strangely called in New Haven. I am too aristocratic for such doings as *help* would make those who live in New Haven endure. Ardently as I am attached to New Haven the plague of *help* will probably always prevent my living there again, for I would not put up with 'the world turned upside down,' and therefore should give offense to their *helpinesses*, and so lead a very uncomfortable life."

From this our suspicion is strengthened that the servant question belongs to no time or country, but is and always has been a perennial and ubiquitous problem.

"*May 11, 1828.* I feel very anxious about you, dear mother. I heard through Mr. Van Rensselaer that you were better, and I hope that you will yet see many good days on earth and be happy in the affection of your children and friends here, before you go, a little before them, to join those in heaven."

While expressing anxiety about his mother's health, he could not have considered her condition critical, for on the 18th of May he writes again: —

"I did hope so to make my arrangements as to have been with you in New Haven yesterday and to-day, but I am so situated as to be unable to leave the city without great detriment to my business. . . . Unless, therefore, there is something of pressing necessity, prudence would dictate to me to take advantage of this season, which has generally been the most profitable to others in the profession, and see if I cannot get my share of something to do. It is a great struggle with me to know what I ought to do. Your situation and that of the family draw me to New Haven; the state of my finances keeps me here. I will come, however, if, on the whole, you think it best."

Again are the records silent as to whether the visit was paid or not, but his anxiety was well founded, for his mother's appointed time had come, and just ten days later, on the 28th of May, 1828, she died at the age of sixty-two.

Thus within the space of three years the hand of death had removed the three beings whom Morse loved best. His mother, while, as we have seen, stern and uncompromising in her Puritan principles, yet possessed the faculty of winning the love as well as the respect of her family and friends. Dr. Todd said of her home: "An orphan myself and never having a home, I have gone away from Dr. Morse's house in tears, feeling that such a home must be more like heaven than anything of which I could conceive."

Mr. Prime, in his biography of Morse, thus pays tribute to her: —

"Two persons more unlike in temperament, it is said, could not have been united in love and marriage than the parents of Morse. The husband was sanguine, impulsive, resolute, regardless of difficulties and danger. She was calm, judicious, cautious, and reflecting. And she, too, had a will of her own. One day she was expressing to one of the parish her intense displeasure with the treatment her husband had received, when Dr. Morse gently laid his hand upon her shoulder and said, 'My dear, you know we must throw the mantle of charity over the imperfections of others.' And she replied with becoming spirit, 'Mr. Morse, charity is not a fool.'"

In the summer of 1828, Morse spent some time in central New York, visiting relatives and painting portraits when the occasion offered. He thus describes a narrow escape from serious injury, or even death, in a letter to his brother Sidney, dated Utica, August 17, 1828: —

"In coming from Whitesboro on Friday I met with an accident and a most narrow escape with my life. The horse, which had been tackled into the wagon, was a vicious horse and had several times run away, to the danger of Mr. Dexter's life and others of the family. I was not aware of this or I should not have consented to go with him, much less to drive him myself.

"I was alone in the wagon with my baggage, and the horse went very well for about a mile, when he gradually quickened his pace and then set out, in spite of all check, on the full run. I kept him in the road, determined to let him run himself tired as the only safe alternative; but just as I came in sight of a piece of the road which had been concealed by an angle, there was a heavy wagon which I must meet so soon that, in order to avoid it, I must give it the whole road.

"This being very narrow, and the ditches and banks on each side very rough, I instantly made up my mind to a serious accident. As well as the velocity of the horse would allow me, however, I kept him on the side, rough as it was, for about a quarter of a mile pretty steadily, expecting, however, to upset every minute; when all at once I saw before me an abrupt, narrow, deep gully into which the wheels on one side were just upon the point of going down. It flashed across me in an instant that, if I could throw the horse down into the ditch, the wheels of the wagon might, perhaps, rest equipoised on each side, and, perhaps, break the horse loose from the wagon.

"I pulled the rein and accomplished the object in part. The sudden plunge of the horse into the gully

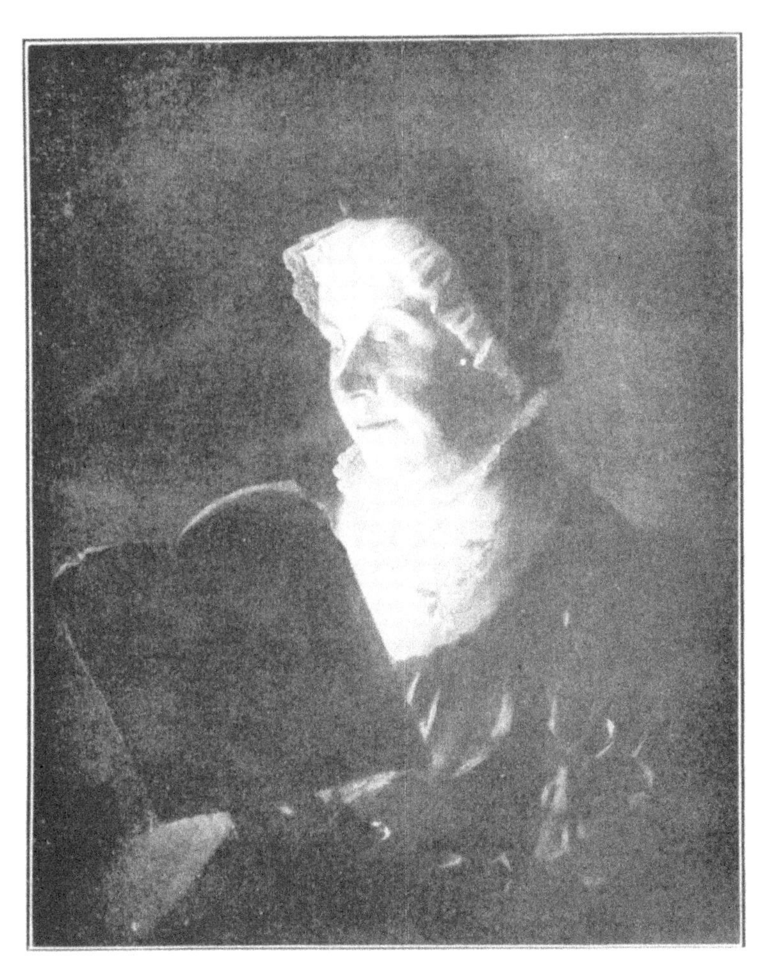

ELIZABETH A. MORSE

Painted by Morse

broke him loose from the wagon, but it at the same time turned one of the fore wheels into the gully, which upset the wagon and threw me forwards at the moment when the horse threw up his heels, just taking off my hat and leaving me in the bottom of the gully. I fell on my left shoulder, and, although muddied from head to foot, I escaped without any injury whatever; I was not even jarred painfully. I found my shoulder a little bruised, my wrist very slightly scratched, and yesterday was a little, and but very little, stiffened in my limbs, and to-day have not the slightest feeling of bruise about me, but think I feel better than I have for a long time. Indeed, my health is entirely restored; the riding and country air have been the means of restoring me. I have great cause of thankfulness for so much mercy and for such special preserving care."

The historian or the biographer who is earnestly desirous of presenting an absolutely truthful picture of men and of events is aided in his task by taking into account the character of the men who have made history. He must ask the question: "Is it conceivable that this man could have acted thus and so under such and such circumstances when his character, as ultimately revealed through the perspective of time, has been established? Could Washington and Lincoln, for example, have been actuated by the motives attributed to them by their enemies?"

Like all men who have become shining marks in the annals of history, Morse could not hope to escape calumny, and in later years he was accused of actions, and motives were imputed to him, which it becomes the duty of his biographer to disprove on the broad ground of moral impossibility.

Among his letters and papers are many rough drafts of thoughts and observations on many subjects, interlined and annotated. Some were afterwards elaborated into letters, articles, or lectures; others seem to have been the thought of the moment, which he yet deemed worth writing down, and which, perhaps better than anything else, reveal the true character of the man.

The following was written by him in pencil on Sunday, September 6, 1829, at Cooperstown, New York: —

"That temptations surround us at every moment is too evident to require proof. If they cease from without they still act upon us from within ourselves, and our most secret thoughts may as surely be drawn from the path of duty by secret temptation, by the admission of evil suggestions, and they will affect our characters as injuriously as those more palpable and tangible temptations that attack our sense.

"This life is a state of discipline; a school in which to form character. There is not an event that comes to our knowledge, not a sentence that we read, not a person with whom we converse, not an act of our lives, in short, not a thought which we conceive, but is acting upon and moulding that character into a shape of good or evil; and, however unconscious we may be of the fact, a thought, casually conceived in the solitariness and silence and darkness of midnight, may so modify and change the current of our future conduct that a blessing or a curse to millions may flow from it.

"All our thoughts are mysteriously connected with good or evil. Their very habits, too, like the habits of our actions, are strengthened by indulgence, and, according as we indulge the evil or the good, our characters will partake of the moral character of each. But actions proceed from thoughts; we act as we think. Why should we, then, so cautiously guard our actions from impropriety while we give a loose rein to our thoughts, which so certainly, sooner or later, produce their fruits in our actions?

"God in his wisdom has separated at various distances sin and the consequence of sin. In some instances we see a sin instantly followed by its fruits, as of revenge by murder. In others we see weeks and months and years, aye, and ages, too, elapse before the fruits of a single act, the result, perhaps, of a single thought, are seen in all their varieties of evil.

"How long ere the fruits of one sin in Paradise will cease to be visible in the moral universe?

"If this reasoning is correct, I shall but cheat myself in preserving a good moral outward appearance to others if every thought of the heart, in the most secret retirement, is not carefully watched and checked and guarded from evil; since the casual indulgence of a single evil thought in secret may be followed, long after that thought is forgotten by me, and when, perhaps, least expected, by overt acts of evil.

"Who, then, shall say that in those pleasures in which we indulge, and which by many are called, and apparently are, innocent, there are not laid the seeds of many a corrupt affection? Who shall say that my innocent indulgence at the card table or at the theatre, were I inclined to visit them, may not produce, if not in me a passion for gaming or for low indulgence, yet in others may encourage these views to their ruin?

"Besides, 'Evil communications corrupt good manners,' and even places less objectionable are studiously to be avoided. The soul is too precious to be thus exposed.

"Where then is our remedy? In Christ alone. 'Cleanse thou me from secret faults. Search me, O God, and know my thoughts; try me and know my ways and see if there is any wicked way in me, and lead me in the way which is everlasting.'"

This is but one of many expressions of a similar character which are to be found in the letters and notes, and which are illuminating.

Morse was now making ready for another trip to Europe. He had hoped, when he returned home in 1815, to stay but a year or two on this side and then to go back and continue his artistic education, which he by no means considered complete, in France and Italy. We have seen how one circumstance after another interfered to prevent the realization of this plan, until now, after the lapse of fourteen years, he found it possible. His wife and his parents were dead; his children were being carefully cared for by relatives, the daughter Susan by her mother's sister, Mrs. Pickering, in Concord, New Hampshire, and the boys by their uncle, Richard C. Morse, who was then happily married and living in the family home in New Haven.

The National Academy of Design was now established on a firm footing and could spare his guiding hand for a few years. He had saved enough money to defray his expenses on a strictly economical basis, but, to make assurance doubly sure, he sought and received commissions from his friends and patrons in America for copies of famous paintings, or for original works of his own, so that he could sail with a clear conscience as regarded his finances.

His friends were uniformly encouraging in furthering his plan, and he received many letters of cordial good wishes and of introduction to prominent men abroad. I shall include the following from John A. Dix, at that time a captain in the army, but afterwards a general, and Governor of New York, who, although he had been an unsuccessful suitor for the hand of Miss Walker, Morse's wife, bore no ill-will towards his rival, but remained his firm friend to the end: —

COOPERSTOWN, 27th October, 1829.

MY DEAR SIR, — I have only time to say that I have been absent in an adjacent county and fear there is not time to procure a letter for you to Mr. Rives before the 1st. I have written to Mr. Van Buren and he will doubtless send you a letter before the 8th. Therefore make arrangements to have it sent after you if you sail on the 1st.

I need not say I shall be very happy to hear from you during your sojournment abroad. Especially tell me what your impressions are when you turn from David's picture with Romulus and Tatius in the foreground, and Paul Veronese's Marriage at Cana directly opposite, at the entrance of the picture gallery in the Louvre.

We are all well and all desire to be remembered. I have only time to add my best wishes for your happiness and prosperity.

Yours truly and constantly,

JOHN A. DIX.

The Mr. Rives mentioned in the letter was at that time our Minister to France, and the Mr. Van Buren was Martin Van Buren, then Secretary of State in President Jackson's Cabinet, and afterwards himself President of the United States.

The following is from the pencilled draft of a letter or the beginning of a diary which was not finished, but ends abruptly: —

"On the 8th November, 1829, I embarked from New York in the ship Napoleon, Captain Smith, for Liverpool. The Napoleon is one of those splendid packets, which have been provided by the enterprise of our merchants, for the accommodation of persons whose business or pleasure requires a visit to Europe or America.

"Precisely at the appointed hour, ten o'clock, the steamboat with the passengers and their baggage left the Whitehall dock for our gallant ship, which was lying to above the city, heading up the North River, careening to the brisk northwest gale, and waiting with apparent impatience for us, like a spirited horse curvetting under the rein of his master, and waiting but his signal to bound away. A few moments brought us to her side, and a few more saw the steamboat leave us, and the sad farewells to relatives and friends, who had thus far accompanied us, were mutually exchanged by the waving of hands and of handkerchiefs. The 'Ready about,' and soon after the 'Mainsail haul' of the pilot were answered by the cheering 'Ho, heave, ho' of the sailors, and, with the fairest wind that ever blew, we fast left the spires and shores of the great city behind us. In two hours we discharged our pilot to the south of Sandy

Hook, with his pocket full of farewell letters to our friends, and then stood on our course for England.

"Four days brought us to the Banks of Newfoundland, one third of our passage. Many of our passengers were sanguine in their anticipations of our making the shortest passage ever known, and, had our subsequent progress been as great as at first, we should doubtless have accomplished the voyage in thirteen days, but calms and head winds for three days on the Banks have frustrated our expectations.

"There is little that is interesting in the incidents of a voyage. The indescribable listlessness of seasickness, the varied state of feeling which changes with the wind and weather, have often been described. These I experienced in all their force. From the time we left the Banks of Newfoundland we had a continued succession of head winds, and when within one fair day's sail of land, we were kept off by severe gales directly ahead for five successive days and nights, during which time the uneasy motion of the ship deprived us all of sleep, except in broken intervals of an half-hour at a time. We neither saw nor spoke any vessel until the evening of the ——, when we descried through the darkness a large vessel on an opposite course from ourselves; we first saw her cabin lights. It was blowing a gale of wind before which we were going on our own course at the rate of eleven miles an hour. It was, of course, impossible to speak her, but, to let her know that she had company on the wide ocean, we threw up a rocket which for splendor of effect surpassed any that I had ever seen on shore. It was thrown from behind the mizzenmast, over which it shot arching its way over the main and foremasts, illuminating every sail and rope, and then diving into the water, piercing the wave, it again shot upwards and vanished in a loud report. To our companion ship the effect must have been very fine.

"The sea is often complained of for its monotony, and yet there is great variety in the appearance of the sea."

Here it ends, but we learn a little more of the voyage and the landing in England from a letter to a cousin in America, written in Liverpool, on December 5, 1829: —

"I arrived safely in England yesterday after a long, but, on the whole, pleasant, passage of twenty-six days. I write you from the inn (the King's Arms Hotel) at which I put up eighteen years ago. This inn is the one at which Professor Silliman stayed when he travelled in England, and which he mentions in his travels. The old Frenchman whom he mentions I well remember when I was here before. I enquired for him and am told he is still living, but I have not seen him.

"There is a large black man, a waiter in the house, who is quite a polished man in his manners, and an elderly white man, with white hair, who looks so respectable and dignified that one feels a little awkward at first in ordering him to do this or that service; and the chambermaids look so venerable and matronly that to ask them for a pitcher of water seems almost rude to them. But I am in a land where domestic servants are the best in the world. No servant aspires to a higher station, but feels a pride in making himself the first in that station. I notice this, for our own country presents a melancholy contrast in this particular."

Here follows a description of the voyage, and he continues: —

"Yesterday we anchored off the Floating Light, sixteen miles from the city, unable to reach the dock on account of the wind, but the post-office steamboat (or steamer, as they call them here) came to us from Liverpool to take the letter-bags, and I with other passengers got on board, and at twelve o'clock I once more placed my foot on English ground.

"The weather is true English weather, thick, smoky, and damp. I can see nothing of the general appearance of the city. The splendid docks, which were building when I was here before, are now completed and extend along the river. They are really splendid; everything about them is solid and substantial, of stone and iron, and on so large a scale.

"I have passed my baggage through the custom-house, and on Monday I proceed on my journey to London through Birmingham and Oxford. Miss Leslie, a sister of my friend Leslie of London, is my *compagnon de voyage*. She is a woman of fine talents and makes my journey less tedious and irksome than it would otherwise be. . . . I have a long journey before me yet ere I reach Rome, where I intended to be by Christmas Day, but my long voyage will probably defeat my intention."

CHAPTER XV

DECEMBER 6, 1829 — FEBRUARY 6, 1830

Journey from Liverpool to London by coach. — Neatness of the cottages. — Trentham Hall. — Stratford-on-Avon. — Oxford. — London. — Charles R. Leslie. — Samuel Rogers. — Seated with Academicians at Royal Academy lecture. — Washington Irving. — Turner. — Leaves London for Dover. — Canterbury Cathedral. — Detained at Dover by bad weather. — Incident of a former visit. — Channel steamer. — Boulogne-sur-Mer. — First impressions of France. — Paris. — The Louvre. — Lafayette. — Cold in Paris. — Continental Sunday. — Leaves Paris for Marseilles in diligence. — Intense cold. — Dijon. — French funeral. — Lyons. — The Hôtel Dieu. —Avignon. — Catholic church services. — Marseilles. — Toulon. — The navy yard and the galley slaves. — Disagreeable experience at an inn. — The Riviera. — Genoa.

MORSE was now thirty-eight years old, in the full vigor of manhood, of a spare but well-knit frame and of a strong constitution. While all his life, and especially in his younger years, he was a sufferer from occasional severe headaches, he never let these interfere with the work on hand, and, by leading a sane and rational life, he escaped all serious illnesses. He was not a total abstainer as regards either wine or tobacco, but was moderate in the use of both; a temperance advocate in the true sense of the word.

His character had now been moulded both by prosperity and adversity. He had known the love of wife and children, and of father and mother, and the cup of domestic happiness had been dashed from his lips. He had experienced the joy of the artist in successful creation, and the bitterness of the sensitive soul irritated by the ignorant, and all but overwhelmed by the struggle for existence. He had felt the supreme joy of swaying an audience by his eloquence, and he had endured with fortitude the carping criticism of the envious. Through it all, through prosperity and through adversity, his hopeful, buoyant nature had triumphed. Prosperity had not spoiled him, and adversity had but served to refine. He felt that he had been given talents which he must utilize to the utmost, that he must be true to himself, and that, above all, he must strive in every way to benefit his fellow men.

This motive we find recurring again and again in his correspondence and in his intimate notes. Not, "What can I do for myself?" but "What can I do for mankind?" Never falsely humble, but, on the contrary, properly proud of his achievements, jealous of his own good name and fame and eager *honestly* to acquire wealth, he yet ever put the public good above his private gain.

He was now again in Europe, the goal of his desires for many years, and he was about to visit the Continent, where he had never been. Paris, with her treasures of art, Italy, the promised land of every artist, lay before him.

We shall miss the many intimate letters to his wife and to his parents, but we shall find others to his brothers and to his friends, perhaps a shade less unreserved, but still giving a clear account of his wanderings, and, from a mass of little notebooks and sketch-books, we can follow him on his pilgrimage and glean some keen observations on the peoples and places visited by him. It must be remembered that this was still the era of the stage-coach and the diligence, and that it took many days to accomplish a journey which is now made in almost the same number of hours.

On Christmas Day, 1829, he begins a letter from Dover to a favorite cousin, Mrs. Margaret Roby, of Utica, New York: —

"When I left Liverpool I took my seat upon the outside of the coach, in order to see as much as possible of the country through which I was to pass. Unfortunately the fog and smoke were so dense that I could see objects but a few yards from the road. Occasionally, indeed, the fog would become less dense, and we could see the fine lawns of the seats of the nobility and gentry, which were scattered on our route, and which still retained their verdure. Now and then the spire and towers of some ancient village church rose out of the leafless trees, beautifully simple in their forms, and sometimes clothed to the very tops with the evergreen ivy. It was severely cold; my eyebrows, hair, cap, and the fur of my cloak were soon coated with frost, but I determined to keep my seat though I suffered some from the cold.

"Their fine natural health, or the frosty weather, gave to the complexions of the peasantry, particularly the females and children, a beautiful rosy bloom. Through all the villages there was the appearance of great comfort and neatness, — a neatness, however, very different from ours. Their nicely thatched cottages bore all the marks of great antiquity, covered with brilliant green moss like velvet, and round the doors and windows were trained some of the many kinds of evergreen vines which abound here. Most of them also had a trim courtyard before their doors, planted with laurel and holly and box, and sometimes a yew cut into some fantastic shape. The whole appearance of the villages was neat and venerable; like some aged matron who, with all her wrinkles, her stooping form, and grey locks, preserves the dignity of cleanliness in her ancient but becoming costume.

"At Trentham we passed one of the seats of the Marquis of Stafford, Trentham Hall. Here the Marquis has a fine gallery of pictures, and among them Allston's famous picture of 'Uriel in the Sun.'

"I slept the first night in Birmingham, which I had no time to see on account of darkness, smoke, and fog: three most inveterate enemies to the seekers of the picturesque and of antiquities. In the morning, before daylight, I resumed my journey towards London. At Stratford-on-Avon I breakfasted, but in such haste as not to be able to visit again the house of Shakespeare's birth, or his tomb. This house, however, I visited when in England before. At Oxford, the city of so many classical recollections, I stopped but a few moments to dine. I was here also when before in England. It is a most splendid city; its spires and domes and towers and pinnacles, rising from amid the trees, give it a magnificent appearance as you approach it.

"Before we reached Oxford we passed through Woodstock and Blenheim, the seat of the Duke of Marlborough, whose splendid estates are at present suffering from the embarrassment of the present Duke, who has ruined his fortunes by his fondness for play.

"Darkness came on after leaving Oxford; I saw nothing until arriving in the vicinity of the great metropolis, which has, for many miles before you enter it, the appearance of a continuous village. We saw the brilliant gas-lights of its streets, and our coach soon joined the throng of vehicles that rattled over its pavements.

I could scarcely realize that I was once more in London after fourteen years' absence.

"My first visit was to my old friend and fellow pupil, Leslie, who seemed overjoyed to see me and has been unremitting in his attentions during my stay in London. Leslie I found, as I expected, in high favor with the highest classes of England's noblemen and literary characters. His reputation is well deserved and will not be ephemeral.

"I received an invitation to breakfast from Samuel Rogers, Esq., the celebrated poet, which I accepted with my friend Leslie. Mr. Rogers is the author of 'Pleasures of Memory,' of 'Italy,' and other poems. He has not the proverbial lot of the poet, — that of being poor, — for he is one of the wealthiest bankers and lives in splendid style. His collection of pictures is very select, chosen by himself with great taste.

"I attended, a few evenings since, the lecture on anatomy at the Royal Academy, where I was introduced to some of the most distinguished artists; to Mr. Shee, the poet and author as well as painter; to Mr. Howard, the secretary of the Academy; to Mr. Hilton, the keeper; to Mr. Stothard, the librarian; and several others. I expected to have met and been introduced to Sir Thomas Lawrence, the president, but he was absent, and I have not had the pleasure of seeing him. I was invited to a seat with the Academicians, as was also Mr. Cole, a member of our Academy in New York. I was gratified in seeing America so well represented in the painters Leslie and Newton. The lecturer also paid, in his lecture, a high compliment to Allston by a deserved panegyric, and by several quotations from his poems, illustrative of principles which he advanced.

"After the lecture I went home to tea with Newton, accompanied by Leslie, where I found our distinguished countryman, Washington Irving, our Secretary of Legation, and W. E. West, another American painter, whose portrait of Lord Byron gave him much celebrity. I passed a very pleasant evening, of course.

"The next day I visited the National Gallery of pictures, as yet but small, but containing some of the finest pictures in England. Among them is the celebrated 'Raising of Lazarus' by Sebastian del Piombo, for which a nobleman of this country offered to the late proprietor sixteen thousand pounds sterling, which sum was refused. I visited also Mr. Turner, the best landscape painter living, and was introduced to him. . . .

"I did not see so much of London or its curiosities as I should have done at another season of the year. The greater part of the time was night — literally night; for, besides being the shortest days of the year (it not being light until eight o'clock and dark again at four), the smoke and fog have been most of the time so dense that darkness has for many days occupied the hours of daylight. . . .

"On the 22d inst., Tuesday, I left London, after having obtained in due form my passports, for the Continent, in company with J. Town, Esq., and N. Jocelyn, Esq., American friends, intending to pass the night at Canterbury, thirty-six miles from London. The day was very unpleasant, very cold, and snowing most of the time. At Blackheath we saw the palace in which the late unfortunate queen of George IV resided. On the heath among the bushes is a low furze with which it is in part covered. There were encamped in their miserable blanket huts a gang of gypsies. No wigwams of the Oneidas ever looked so comfortless. On the road we overtook a gypsy girl with a child in her arms, both having the stamp of that singular race strongly marked upon their features; black hair and sparkling black eyes, with a nut-brown complexion and cheeks of russet red, and not without a shrewd intelligence in their expression.

"At about nine o'clock we arrived at the Guildhall Tavern in the celebrated and ancient city of Canterbury. Early in the morning, as soon as we had breakfasted, we visited the superb cathedral. This stupendous pile is one of the most distinguished Gothic structures in the world. It is not only interesting from its imposing style of architecture, but from its numerous historical associations. The first glimpse we caught of it was through and over a rich, decayed gateway to the enclosure of the cathedral grounds. After passing the gate the vast pile — with its three great towers and innumerable turrets, and pinnacles, and buttresses, and arches, and painted windows — rose in majesty before us. The grand centre tower, covered with a grey moss, seemed like an immense mass of the Palisades, struck out with all its regular irregularity, and placed above the surrounding masses of the same grey rocks. The bell of the great tower was tolling for morning service, and yet so distant, from its height, that it was scarcely heard upon the pavement below.

"We entered the door of one of the towers and came immediately into the nave of the church. The effect of the long aisles and towering, clustered pillars and richly carved screens of a Gothic church upon the imagination can scarcely be described — the emotion is that of awe.

"A short procession was quickly passing up the steps of the choir, consisting of the beadle, or some such officer, with his wand of office, followed by ten boys in white surplices. Behind these were the prebendaries and other officers of the church; one thin and pale, another portly and round, with powdered hair and sleepy, dull, heavy expression of face, much like the face that Hogarth has chosen for the 'Preacher to his Sleepy Congregation.' This personage we afterward heard was Lord Nelson, the brother of the celebrated Nelson and the heir to his title.

"The service was read in a hurried and commonplace manner to about thirty individuals, most of whom seemed to be the necessary assistants at the ceremonies. The effect of the voices in the responses and the chanting of the boys, reverberating through the aisles and arches and recesses of the church, was peculiarly imposing, but, when the great organ struck in, the emotion of grandeur was carried to its height, — I say nothing of devotion. I did not pretend on this occasion to join in it; I own that my thoughts as well as my eyes were roaming to other objects, and gathering around me the thousand recollections of scenic splendor, of terror, of bigotry, and superstition which were acted in sight of the very walls by which I was surrounded. Here the murder of Thomas à Becket was perpetrated; there was his miracle-working shrine, visited by pilgrims from all parts of Christendom, and enriched with the most costly jewels that the wealth of princes could purchase and lavish upon it; the very steps, worn into deep cavities by the knees of the devotees as they approached the shrine, were ascended by us. There stood the tomb of Henry

IV and his queen; and here was the tomb of Edward, the Black Prince, with a bronze figure of the prince, richly embossed and enamelled, reclining upon the top, and over the canopy were suspended the surcoat and casque, the gloves of mail and shield, with which he was accoutred when he fought the famous battle of Crécy. There also stood the marble chair in which the Saxon kings were crowned, and in which, with the natural desire that all seemed to have in such cases, I could not avoid seating myself. From this chair, placed at one end of the nave, is seen to best advantage the length of the church, five hundred feet in extent.

"After the service I visited more at leisure the tombs and other curiosities of the church. The precise spot on which Archbishop Becket was murdered is shown, but the spot on which his head fell on the pavement was cut out as a relic and sent to Rome, and the place filled in with a fresh piece of stone, about five inches square. . . .

"In the afternoon we left Canterbury and proceeded to Dover, intending to embark the next morning (Thursday, December 24) for Calais or Boulogne in the steamer. The weather, however, was very unpromising in the morning, being thick and foggy and apparently preparing for a storm. We therefore made up our minds to stay, hoping the next day would be more favorable; but Friday, Christmas Day, came with a most violent northeast gale and snowstorm. Saturday the 26th, Sunday the 27th, and, at this moment, Monday the 28th, the storm is more violent than ever, the streets are clogged with snow, and we are thus embargoed completely for we know not how long a time to come.

"Notwithstanding the severity of the weather on

Thursday, we all ventured out through the wind and snow to visit Dover Castle, situated upon the bleak cliffs to the north of the town. . . .

"The castle, with its various towers and walls and outworks, has been the constant care of the Government for ages. Here are the remains of every age from the time of the Romans to the present. About the centre of the enclosure stand two ancient ruins, the one a tower built by the Romans, thirty-six years after Christ, and the other a rude church built by the Saxons in the sixth century. Other remains of towers and walls indicate the various kinds of defensive and offensive war in different ages, from the time when the round or square tower, with its loopholes for the archers and crossbowmen, and gates secured by heavy portcullis, were a substantial defence, down to the present time, when the bastion of regular sides advances from the glacis, mounted with modern ordnance, keeping at a greater distance the hostile besiegers.

"Through the glacis in various parts are sally-ports, from one of which, opening towards the road to Ramsgate, I well remember seeing a corporal's guard issue, about fifteen years ago, to take possession of me and my sketch-book, as I sat under a hedge at some distance to sketch the picturesque towers of this castle. Somewhat suspicious of their intentions, I left my retreat, and, by a circuitous route into the town, made my escape; not, however, without ascertaining from behind a distant hedge that I was actually the object of their expedition. They went to the spot where I had been sitting, made a short search, and then returned to the castle through the same sally-port.

"At that time (a time of war not only with France but America also) the strictest watch was kept, and to have been caught making the slightest sketch of a fortification would have subjected me to much trouble. Times are now changed, and had Jack Frost (the only commander of rigor now at the castle) permitted, I might have sketched any part of the interior or exterior."

"*Boulogne-sur-Mer, France, December 29, 1829.* This morning at ten o'clock, after our tedious detention, we embarked from Dover in a steamer for this place instead of Calais. I mentioned the steamer, but, cousin, if you have formed any idea of elegance, or comfort, or speed in connection with the name of steamer from seeing our fine steamboats, and have imagined that English or French boats are superior to ours, you may as well be undeceived. I know of no description of packet-boats in our waters bad enough to convey the idea. They are small, black, dirty, confined things, which would be suffered to rot at the wharves for want of the least custom from the lowest in our country. You may judge of the extent of the accommodations when I tell you that there is in them but one cabin, six feet six inches high, fourteen feet long, eleven feet wide, containing eight berths.

"Our passage was, fortunately, short, and we arrived in the dominions of 'His Most Christian Majesty' Charles X at five o'clock. The transition from a country where one's own language is spoken to one where the accents are strange; from a country where the manners and habits are somewhat allied to our own to one where everything is different, even to the most trifling article of dress, is very striking on landing after so short an interval from England to France.

"The pier-head at our landing was filled with human beings in strange costume, from the grey *surtout* and belt of the *gendarmes* to the broad twilled and curiously plaited caps of the masculine women; which latter beings, by the way, are the licensed porters of baggage to the custom-house."

"*Paris, January 7, 1830.* Here have I been in this great capital of the Continent since the first day of the year. I shall remember my first visit to Paris from the circumstance that, at the dawn of the day of the new year, we passed the Porte Saint-Denis into the narrow and dirty streets of the great metropolis.

"The Louvre was the first object we visited. Our passports obtained us ready admittance, and, although our fingers and feet were almost frozen, we yet lingered three hours in the grand gallery of pictures. Indeed, it is a long walk simply to pass up and down the long hall, the end of which from the opposite end is scarcely visible, but is lost in the mist of distance. On the walls are twelve hundred and fifty of some of the *chefs d'œuvre* of painting. Here I have marked out several which I shall copy on my return from Italy.

"I have my residence at present at the Hôtel de Lille, which is situated very conveniently in the midst of all the most interesting objects of curiosity to a stranger in Paris, — the palace of the Tuileries, the Palais Royal, the Bibliothèque Royale, or Royal Library, and numerous other places, all within a few paces of us. On New Year's Day the equipages of the nobility and foreign ambassadors, etc., who paid their respects to the King and the Duke of Orléans, made considerable display in the Place du Carrousel and in the court of the Tuileries.

"At an exhibition of manufactures of porcelain, tapestry, etc., in the Louvre, where were some of the most superb specimens of art in the world in these articles, we also saw the Duchesse de Berri. She is the mother of the little Duc de Bordeaux, who, you know, is the heir apparent to the crown of France. She was simply habited in a blue pelisse and blue bonnet, and would not be distinguished in her appearance from the crowd except by her attendants in livery.

"I cannot close, however, without telling you what a delightful evening I passed evening before last at General Lafayette's. He had a soirée on that night at which there were a number of Americans. When I went in he instantly recognized me; took me by both hands; said he was expecting to see me in France, having read in the American papers that I had embarked. He met me apparently with great cordiality, then introduced me to each of his family, to his daughters, to Madame Lasterie and her two daughters (very pretty girls) and to Madame Rémusat,[1] and two daughters of his son, G. W. Lafayette, also very accomplished and beautiful girls. The General inquired how long I intended to stay in France, and pressed me to come and pass some time at La Grange when I returned from Italy. General Lafayette looks very well and seems to have the respect of all the best men in France. At his soirée I saw the celebrated Benjamin Constant, one of the most distinguished of the Liberal party in France. He is tall and thin with a very fair, white complexion, and long white, silken hair, moving with all the vigor of a young man."

[1] This was not, of course, the famous Madame de Rémusat; probably her daughter-in-law.

In a letter to his brothers written on the same day, January 7th, he says: —

"If I went no farther and should now return, what I have already seen and studied would be worth to me all the trouble and expense thus far incurred. I am more and more satisfied that my expedition was wisely planned.

"You cannot conceive how the cold is felt in Paris, and, indeed, in all France. Not that their climate is so intensely cold as ours, but their provision against the cold is so bad. Fuel is excessively high; their fireplaces constructed on the worst possible plan, looking like great ovens dug four or five feet into the wall, wasting a vast deal of heat; and then the doors and windows are far from tight; so that, altogether, Paris in winter is not the most comfortable place in the world.

"Mr. Town and I, and probably Mr. Jocelyn, set out for Italy on Monday by the way of Châlons-sur-Saone, Lyons, Avignon, and Nice. I long to get to Rome and Naples that I may commence to paint in a warm climate, and so keep warm weather with me to France again. . . .

"I don't know what to do about writing letters for the 'Journal of Commerce.' I fear it will consume more of my time than the thing is worth, and will be such a hindrance to my professional studies that I must, on the whole, give up the thought of it. My time here is worth a guinea a minute in the way of my profession. I could undoubtedly write some interesting letters for them, but I do not feel the same ease in writing for the public that I do in writing to a friend, and, in correcting my language for the press, I feel that it is going to consume more of my time than I can spare. I will write if I can, but they must not expect it, for I find my pen and pencil are enemies to each other. I must write less and paint more. My advantages for study never appeared so great, and I never felt so ardent a desire to improve them."

Morse spent about two weeks in Paris visiting churches, picture galleries, palaces, and other show places. He finds the giraffe or camelopard the most interesting animal at the Jardin des Plantes, and he dislikes a ceiling painted by Gros: "It is allegorical, which is a class of painting I detest." He deplores the Continental Sunday: "Oh! that we appreciated in America the value of our Sabbath; a Sabbath of rest from labor; a Sabbath of moral and religious instruction; a Sabbath the greatest barrier to those floods of immorality which have in times past deluged this devoted country in blood, and will again do it unless the Sabbath gains its ascendancy once more."

From an undated and unfinished draft of a letter to his cousin, Mrs. Roby, we learn something of his journey from Paris to Rome, or rather of the first part of it: —

"I wrote you from Paris giving you an account of my travels to that city, and I now improve the first moments of leisure since to continue my journal. After getting our passports signed by at least half a dozen ambassadors preparatory to our long journey, we left Paris on Wednesday, January 13, at eight o'clock, for Dijon, in the diligence. The weather was very cold, and we travelled through a very uninteresting country. It seemed like a frozen ocean, the road being over an immense plain unbroken by trees or fences.

"We stopped a few moments at Melun, at Joigny and

Tonnerre, which latter place was quite pretty with a fine-looking Gothic church. We found the villages from Paris thus far much neater and in better style than those on the road from Boulogne.

"Our company consisted of Mr. Town, of New York, Mr. Jocelyn, of New Haven, a very pretty Frenchwoman, and myself. The Frenchwoman was quite a character; she could not talk English nor could we talk French, and yet we were talking all the time, and were able to understand and be understood.

"At four o'clock the next morning we *dined!!* at Montbar, which place we entered after much detention by the snow. It was so deep that we were repeatedly stopped for some time. At a picturesque little village, called Val de Luzon, where we changed horses, the country began to assume a different character. It now became mountainous, and, had the season been propitious, many beautiful scenes for the pencil would have presented themselves. As it was, the forms of the mountains and the deep valleys, with villages snugly situated at the bottom, were grateful to the eye amidst the white shroud which everywhere covered the landscape. We could but now and then catch a glimpse of the scenery through our coach window by thawing a place in the thickly covered glass, which was so plated with the arborescent frost as not to yield to the warmth of the sun at midday.

"We arrived at Dijon at nine o'clock on Saturday evening, after three days and two nights of fatiguing riding. The diligence is, on the whole, a comfortable carriage for travelling. I can scarcely give you any idea of its construction; it is so unlike in many respects to our stage-coach. It is three carriage-bodies together upon one set of wheels. The forward part is called the *coupé*, which holds but three persons, and, from having windows in front so that the country is seen as you travel, is the most expensive. The middle carriage is the largest, capable of holding six persons, and is called the *intérieur*. The other, called the *derrière*, is the cheapest, but is generally filled with low people. The *intérieur* is so large and so well cushioned that it is easy to sleep in it ordinarily, and, had it not been for the sudden stops occasioned by the clogging of the wheels in the snow, we should have had very good rest; but the discordant music made by the wheels as they ground the frozen snow, sounding like innumerable instruments, mostly discordant, but now and then concordant, prevented our sound sleep.

"The cold we found as severe as any I have usually experienced in America. The snow is as deep upon the hills, being piled up on each side of the road five or six feet high. The water in our pitchers froze by the fireside, and the glass on the windows, even in rooms comfortably warmed, was encrusted with arborescent frost. The floors, too, of all the rooms are paved with bricks or tiles, and, although comfortable in summer, are far from desirable in such a winter.

"At Dijon we stopped over the Sabbath, for the double purpose of avoiding travelling on that day and from really needing a day of rest. On Sunday morning we enquired of our landlord, Mons. Ripart, of the Hôtel du Parc, for a Protestant church, and were informed that there was not any in the place. We learned, however, afterwards that there was one, but too late to profit by the information. We walked out in the cold to find some church, and, entering a large, irregular Gothic structure, much out of repair, we pressed towards the altar where the funeral service of the Catholic Church was performing over a corpse which lay before it. The priests, seven or eight in number, were in the midst of their ceremonies. They had their hair shorn close in front, but left long behind and at the sides, and powdered, and, while walking, covered partially with a small, black, pyramidal velvet cap with a tuft at the top. While singing the service they held long, lighted wax tapers in their hands. There was much ceremony, but scarcely anything that was imposing; its heartlessness was so apparent, especially in the conduct of some of the assistants, that it seemed a solemn mockery. One in particular, who seemed to pride himself on the manner in which he vociferated 'Amen,' was casting his eyes among the crowd, winking and laughing at various persons, and, from the extravagance of his manners, bawling out most irreverently and closing by laughing, I wondered that he was not perceived and rebuked by the priests.

"As the procession left the church it was headed by an officer bearing a pontoon;[1] then one bearing the silver crucifix; then eight or ten boys with lighted wax tapers by the side of the corpse; then followed the priests, six or eight in number, and then the relatives and friends of the deceased. At the grave the priests and assistants chanted a moment, the coffin was lowered, the earth thrown upon it, and then an elder priest muttered something over the grave, and, with an instrument consisting

[1] This must be a mistake.

of a silver ball with a small handle, made the sign of the cross over the body, which ceremony was repeated by each one in the procession, to whom in succession the instrument was handed.

"There were, indeed, two or three real mourners. One young man in particular, to whom the female might have been related as wife or sister, showed all the signs of heartfelt grief. It did not break out into extravagant gesture or loud cries, but the tears, as they flowed down his manly face, seemed to be forced out by the agony within, which he in vain endeavored to suppress. The struggle to restrain them was manifest, and, as he made the sign of the cross at the grave in his turn, the feebleness with which he performed the ceremony showed that the anguish of his heart had almost overcome his physical strength. I longed to speak to him and to sympathize with him, but my ignorance of the language of his country locked me out from any such purpose. . . .

"Accustomed to the proper and orderly manner of keeping the Sabbath so universal in our country, there are many things that will strike an American not only as singular but disgusting. While in Paris we found it to be customary, not only on week days but also on the Sabbath, to have musicians introduced towards the close of dinner, who play and sing all kinds of songs. We supposed that this custom was a peculiarity of the capital, but this day after dinner a hand-organ played waltzes and songs, and, as if this were not enough, a performer on the guitar succeeded, playing songs, while two or three persons with long cards filled with specimens of natural history — lobsters, crabs, and shells of various kinds — were busy in displaying their handiwork to us, and each concluded his part of the ceremony by presenting a little cup for a contribution."

The letter ends here, and, as I have found but few more of that year, we must depend on his hurriedly written notebooks for a further record of his wanderings.

Leaving Dijon on January 18, Morse and his companions continued their journey through Châlons-sur-Saone, to Macon and Lyons, which they reached late at night. The next two days were spent in viewing the sights of Lyons, which are described at length in his journal. Most of these notes I shall omit. Descriptions of places and of scenery are generally tiresome, except to the authors of them, and I shall transcribe only such portions as have a more than ordinary personal or historic interest. For instance the following entry is characteristic of Morse's simple religious faith: —

"From the Musée we went to the Hôtel Dieu, a hospital on a magnificent and liberal scale. The apartments for the sick were commodiously and neatly arranged. In one of them were two hundred and twelve cots, all of which showed a pale or fevered face upon the pillow. The attendants were women called 'Sisters of Charity,' who have a peculiar costume. These are benevolent women who (some of them of rank and wealth) devote themselves to ministering to the comfort and necessities of the wretched.

"Benevolence is a trait peculiarly feminine. It is seen among women in all countries and all religions, and although true religion sets out this jewel in the greatest beauty, yet superstition and false religions cannot entirely destroy its lustre. It seems to be one of those virtues permitted in a special manner by the Father of all good to survive the ruins of sin on earth, and to withstand the attacks of Satan in his attempts on the happiness of man; and to woman in a marked manner He has confided the keeping of this virtue. She was first in the transgression but last at the cross."

Leaving Lyons at four o'clock on the morning of the 22d, they journeyed slowly towards Avignon, delayed by the condition of the roads covered by an unusual fall of snow which was now melting under the breath of a warm breeze from the south. On the way they pass "between the two hills a telegraph making signals." This was, of course, a semaphore by means of which visual signals were made.

Reaching Avignon on the night of the 23d, they went the next day, which was Sunday, in search of a Protestant church, but none was to be found in this ancient city of the Popes, so they followed a fine military band to the church of St. Agricola and attended the services there, the band participating and making most glorious music.

Morse, with his Puritan background and training, was not much edified by the ritual of the Catholic Church, and, after describing it, he adds: —

"I looked around the church to ascertain what was the effect upon the multitude assembled. The females, kneeling in their chairs, many with their prayer-books reading during the whole ceremony, seemed part of the time engaged in devotional exercises. Far be it from me to say there were not some who were actually devout, hard as it is to conceive of such a thing; but this I will say, that everything around them, instead of aiding devotion, was calculated entirely to destroy it. The imagination was addressed by every avenue; music and painting pressed into the service of — not religion but the contrary — led the mind away from the contemplation of all that is practical in religion to the charms of mere sense. No instruction was imparted; none seems ever to be intended. What but ignorance can be expected when such a system prevails? . . .

"Last evening we were delighted with some exquisite sacred music, sung apparently by men's voices only, and slowly passing under our windows. The whole effect was enchanting; the various parts were so harmoniously adapted and the taste with which these unknown minstrels strengthened and softened their tones gave us, with the recollection of the music at the church, which we had heard in the morning, a high idea of the musical talent of this part of the world. We have observed more beautiful faces among the women in a single day in Avignon than during the two weeks we were in Paris."

After a three days' rest in Avignon, visiting the palace of the Popes and other objects of interest, and being quite charmed with the city as a whole and with the Hôtel de l'Europe in particular, the little party left for Marseilles by way of Aix. The air grows balmier as they near the Mediterranean, and they are delighted with the vineyards and the olive groves. The first sight of the blue sea and of the beautiful harbor of Marseilles rouses the enthusiasm of the artist, and some days are spent in exploring the city.

The journal continues: —

"*Thursday, January 28.* Took our seats in the Malle Poste for Toulon and experienced one of those vexations in delay which travellers must expect sometimes to find. We had been told by the officer that we must be ready to go at one o'clock. We were, of course, ready at that time, but not only were we not called at one, but we waited in suspense until six o'clock in the evening before we were called, and before we left the city it was seven o'clock; thus consuming a half-day of daylight which we had promised ourselves to see the scenery, and bringing all our travelling in the night, which we wished specially to avoid. Besides this, we found ourselves in a little, miserable, jolting vehicle that did not, like the diligence, suffer us to sleep.

"Thus we left Marseilles, pursuing our way through what seemed to us a wild country, with many a dark ravine on our roadside and impending cliffs above us; a safe resort for bandits to annoy the traveller if they felt disposed."

At Toulon they visited the arsenal and navy yard.

"We saw many ships of all classes in various states of equipment, and every indication, from the activity which pervaded every department, that great attention is paying by the French to their marine. Their ships have not the neatness of ours; there seems to be a great deal of ornament, and such as I should suppose was worse than useless in a ship of war.

"We noticed the galley slaves at work; they had a peculiar dress to mark them. They were dressed in red frocks with the letters 'G a l' stamped on each side of the back, as they were also on their pantaloons. The worst sort, those who had committed murder, had been shipped lately to Brest. Those who had been convicted twice had on a green cap; those who were ordinary criminals had on a red cap; and those who were least criminal, a blue cap.

"A great mortality was prevailing among them. There are about five hundred at this place, and I was told by the sentinel that twenty-two had been buried yesterday. Three bodies were carried out whilst we were in the yard. We, of course, did not linger in the vicinity of the hospitals. . . .

"On Saturday, January 30, we left Toulon in a *voiture* or private carriage, the public conveyances towards Italy being now uncertain, inconvenient, and expensive. There were five of us and we made an agreement in writing with a *vetturino* to carry us to Nice, the first city in Italy, for twenty-seven francs each, the same as the fare in the diligence, to which place he agreed to take us in two days and a half. Of course necessity obliges us in this instance to travel on the Sabbath, which we tried every means in our power to avoid.

"At twelve we stopped at the village of Cuers, an obscure, dirty place, and stopped at an inn called 'La Croix d'Or' for breakfast. We here met with the first gross imposition in charges that occurred to us in France. Our *déjeuner* for five consisted of three cups of miserable coffee, without milk or butter; a piece of beef stewed with olives for two; mutton chops for five; eggs for five; some cheese, and a meagre dessert of raisins, hazel nuts, and olives, with a bottle of sour *vin ordinaire;* and for this we were charged fifteen francs, or three francs each, while at the best hotels in Paris, and in all the cities through which we passed, we had double the quantity of fare, and of the best kind, for two francs and sometimes for one and one half francs. All parleying with the extortionate landlord had only the effect of making him more positive and even insolent; and when we at last threw him the money to avoid further detention, he told us to mark his house, and, with the face of a demon, told us we should never enter his house again. We can easily bear our punishment. As we resumed our journey we were saluted with a shower of stones."

The journal continues and tells of the slow progress along the Riviera, through Cannes, which was then but an unimportant village; Nice, at that time belonging to Italy, and where they saw in the cathedral Charles Felix, King of Sardinia. It took them many days to climb up and down the rugged road over the mountains, while now the traveller is whisked under and around the same mountains in a few hours.

"At eleven we had attained a height of at least two thousand feet and the precipices became frightful, sweeping down into long ravines to the very edge of the sea; and then the road would wind at the edge of the precipice two or three thousand feet deep. Such scenes pass so rapidly it is impossible to make note of them.

"From the heights on which La Turbia stands, with its dilapidated walls, we see the beautiful city of Monaco, on a tongue of land extending into the sea."

The great gambling establishment of Monte Carlo did not invade this beautiful spot until many years later, in 1856.

The travellers stopped for a few hours at Mentone,— "a beautiful place for an artist," — passed the night at San Remo, and, sauntering thus leisurely along the beautiful Riviera, arrived in Genoa on the 6th of February.

JEREMIAH EVARTS

From a portrait painted by Morse owned by Sherman Evarts, Esq.

CHAPTER XVI

FEBRUARY 6, 1830 — JUNE 15, 1830

Serra Palace in Genoa. — Starts for Rome. — Rain in the mountains. — A brigand. — Carrara. — First mention of a railroad. — Pisa. — The leaning tower. — Rome at last. — Begins copying at once. — Notebooks. — Ceremonies at the Vatican. — Pope Pius VIII. — Academy of St. Luke's. — St. Peter's. — Chiesa Nuova. — Painting at the Vatican. — Beggar monks. — *Festa* of the Annunciation. — Soirée at Palazzo Simbaldi. — Passion Sunday. — Horace Vernet. — Lying in state of a cardinal. — *Miserere* at Sistine Chapel. — Holy Thursday at St. Peter's. — Third cardinal dies. — Meets Thorwaldsen at Signor Persianis's. — Manners of English, French, and Americans. — Landi's pictures. — Funeral of a young girl. — Trip to Tivoli, Subiaco. — Procession of the *Corpus Domini*. — Disagreeable experience.

THE enthusiastic artist was now in Italy, the land of his dreams, and his notebooks are filled with short comments or longer descriptions of churches, palaces, and pictures in Genoa and in the other towns through which he passed on his way to Rome, or with pen-pictures of the wild country through which he and his fellow travellers journeyed.

In Genoa, where he stopped several days, he was delighted with the palaces and churches, and yet he found material for criticism: —

"The next place of interest was the Serra Palace, now inhabited by one of that family, who, we understood, was insane. After stopping a moment in the anteroom, the ceiling of which is painted in fresco by Somnio, we were ushered into the room called the most splendid in Europe, and, if carving and gilding and mirrors and chandeliers and costly colors can make a splendid room, this is certainly that room. The chandeliers and mirrored sides are so arranged as to create the illusion that the room is of indefinite extent. To me it appeared, on the whole, tawdry, seeing it in broad daylight. In the evening, when the chandeliers are lighted, I have no doubt of its being a most gorgeous exhibition, but, like some showy belle dressed and painted for evening effect, the daylight turns her gold into tinsel and her bloom into rouge.

"After having stayed nearly four days in Genoa, and after having made arrangements with our honest *vetturino*, Dominique, to take us to Rome, stopping at various places on the way long enough to see them, we retired late to bed to prepare for our journey in the morning.

"On Wednesday morning, February 10, we rose at five o'clock, and, after breakfast of coffee, etc., we set out at six on our journey towards Rome."

I shall not follow them every step of the way, but shall select only the more personal entries in the diary.

"A little after eleven o'clock we stopped at a single house upon a high hill overlooking the sea, to breakfast. It has the imposing title of 'Locanda della Gran Bretagna.' We expected little and got less, and had a specimen of the bad faith of these people. We enquired the price of our *déjeuner* before we ordered it, which is always necessary. We were told one franc each, but after our breakfast, we were told one and a half each, and no talking with the landlord would alter his determination to demand his price. There is no remedy for travellers; they must pay or be delayed.

"At one o'clock we left this hole of a place, where we were more beset with beggars and spongers than at any place since we had been in Italy."

Stopping overnight at Sestri, they set out again on the 11th at five o'clock in the morning: —

"It was as dark as the moon, obscured by thick clouds, would allow it to be, and, as we left the courtyard of the inn, it began to rain violently. Our road lay over precipitous mountains away from the shore, and the scenery became wild and grand. As the day dawned we found ourselves in the midst of stupendous mountains rising in cones from the valleys below. Deep basins were formed at the bottom by the meeting of the long slopes; clouds were seen far below us, some wasting away as they sailed over the steeps, and some gathering denseness as they were detained by the cold, snowy peaks which shot up beyond. Now and then a winding stream glittered at the bottom of some deep ravine amidst the darkness around it, and occasionally a light from the cottage of some peasant glimmered like a star through the clouds.

"As we labored up the steep ascent little brawling cascades without number, from the heights far above us, in milky streams, gathering power from innumerable rills, dashed at our feet, and, passing down through the artificial passages beneath the road, swept down into the valleys in torrents, and swelling the rivers, whose broad beds were seen through the openings, rushed with irresistible power to the sea.

"We found, from the violence of the storm, that the road was heavy and much injured in some parts by the washing down of rocks from the heights. Some of great size lay at the sides recently thrown down, and now and then one of some hundred pounds' weight was found in the middle of the road.

"We continued to ascend about four hours until we came again from a region of summer into the region of snow, and the height from the sea was greater than we had at any time previously attained. The scenery around us, too, was wilder and more sterile. The Apennines here are very grand, assuming every variety of shape and color. Long slopes of clay color were interlocked with dark browns sprinkled with golden yellow; slate blue and grey, mixed with greens and purples, and the pure, deep ultramarine blue of distant peaks finished the background."

After breakfasting at Borghetto at a miserable inn, where they were much annoyed by beggars of all descriptions, they continued their journey through much the same character of country for the rest of the day, and towards dark they met with a slight adventure: —

"Our road was down a steep declivity winding much in the same way as at Finale. Precipices were at the side without a protecting barrier, and we felt some uneasiness at our situation, which was not decreased by suddenly finding our coach stopped and a man on horseback (or rather muleback) stopping by the side of the coach. It was but for a moment; our *vetturino* authoritatively ordered him to pass on, which he did with a '*buona sera,*' and we never parted with a companion more gladly. From all the circumstances attending it we were inclined to believe that he had some design upon us, but, finding us so numerous, thought it best not to run the risk."

Spezia was their resting-place for that night, and, after an early start the next morning, they reached the banks of the Vara at nine o'clock.

"We had a singular time in passing the river in a boat. Many women of the lower orders crossed at the same time. The boat being unable to approach the shore, we were obliged to ride papoose-back upon the shoulders of the brawny watermen for some little distance; but what amused us much was the perfect *sang-froid* with which the women, with their bare legs, held up their clothes above the knees and waded to the boat before us. . . .

"At half-past twelve we came in sight of Carrara. This place we went out of our course to see, and at one o'clock entered the celebrated village, prettily situated in a valley at the base of stupendous mountains. A deep ravine above the village contains the principal quarries of most exquisite marbles for which this place has for so many ages been famous. The clouds obscuring the highest peaks, and ascending from the valleys like smoke from the craters of many volcanoes, gave additional grandeur to a scene by nature so grand in itself.

"After stopping at the Hôtel de Nouvelle Paros, which we found a miserable inn with bad wine, scanty fare and high charges, we took a hasty breakfast, and procuring a guide we walked out to see the curiosities of the place. It rained hard and the road was excessively bad, sometimes almost ankle-deep in mud. Notwithstanding the forbidding weather and bad road, we labored up the deep ravine on the sides of which the excavations are made. Dark peaks frowned above us capped with clouds and snow; white patches midway the sides showed the veins of the marble, and immense heaps of detritus, the accumulation of ages, mountains themselves, sloped down on each side like masses of piled ice to the very edge of the road. The road itself, white with the material of which it is made, was composed of loose pieces of the white marble of every size. . . . Continuing the ascent by the side of a milky stream, which rushed down its rocky bed, and which here and there was diverted off into aqueducts to the various mills, we were pointed to the top of a high hill by the roadside where was the entrance to a celebrated grotto, and at the base close by, a cavern protected a beautiful, clear, crystal fountain, which gushed from up the bottom forming a liquid, transparent floor, and then glided to mingle its pure, unsullied waters with the cloudy stream that rushed by it.

"Climbing over piles of rock like refined sugar and passing several wagons carrying heavy blocks down the road, we arrived at the mouth of the principal quarry where the purest statuary marble is obtained. I could not but think how many exquisite statues here lay entombed for ages, till genius, at various times, called them from their slumbers and bid them live. . . .

"On our return we again passed the wagons laden with blocks, and mules with slabs on each side sometimes like the roof of a house over the mule. . . . The wagons and oxen deserve notice. The former are very badly constructed; they are strong, but the wheels are small, in diameter about two feet and but about three inches wide, so sharp that the roads must suffer from them. The oxen are small and, without exception, mouse-colored. The driver, and there is usually one to each pair, sits on the yoke between them, and, like the oarsman of a boat, with his back towards the point towards which he is going. Two huge blocks were chained upon one of these wagons, and behind, dragging upon the ground by a chain, was another. Three yoke of these small oxen, apparently without fatigue, drew the load thus constructed over this wretched road. An enterprising company of Americans or English, by the construction of a railroad, which is more practicable than a canal, but which latter might be constructed, would, I should think, give great activity to the operations here and make it very profitable to themselves."

It is rather curious to note that this is the first mention of a railroad made by Morse in his notes or letters, although he was evidently aware of the experiments which were being made at that time both in Europe and America, and these must have been of great interest to him. It is also well to bear in mind that the great development of transportation by rail could not occur until the invention of the telegraph had made it possible to send signals ahead, and, in other ways, to control the movement of traffic. At the present day the railroad at Carrara, which Morse saw in his visions of the future, has been built, but the ox teams are also still used, and linger as a reminder of more primitive days.

Continuing their journey, the travellers spent the night at Lucca, and in the morning explored the town, which they found most interesting as well as neat and clean. Leaving Lucca, "with much reluctance," on the 13th, the journal continues: —

"At half-past five, at sunset, Pisa with its leaning tower (the *duomo* of the cathedral and that of the baptistery being the principal objects in the view), was seen across the plain before us. Towards the west was a long line of horizon, unbroken, except here and there by a low-roofed tower or the little pyramidal spire of a village church. To the southeast the plain stretched away to the base of distant blue mountains, and to the east and the north the rude peaks through which we had travelled, their cold tops tinged with a warmer glow, glittered beyond the deep brown slopes, which were more advanced and confining the plain to narrower limits."

They found the Hôtel Royal de l'Hussar an excellent inn, and, the next day being Sunday, they attended an English service and heard an excellent sermon by the Reverend Mr. Ford, an Englishman.

"In the evening we walked to the famous leaning tower, the cathedral, the baptistery, and Campo Santo, which are clustered together in the northern part of the city. In going there we went some distance along the quay, which was filled with carriages and pedestrians, among whom were many masques and fancy dresses of the most grotesque kind. It is the season of Carnival, and all these fooleries are permitted at this time. We merely glanced at the exterior of the celebrated buildings, leaving till to-morrow a more thorough examination."

"*Monday, February 15.* We rose early and went again to the leaning tower and its associated buildings. The tower, which is the *campanile* of the cathedral and is about one hundred and ninety feet high, leans from its perpendicular thirteen feet. We ascended to the top by a winding staircase. One ascending feels the inclination every step he takes, and, when he reaches the top and perceives that that which should be horizontal is an inclined plane, the sensation is truly startling. It is difficult to persuade one's self that the tower is not actually falling, and I could not but imagine at intervals that it moved, reasoning myself momentarily into security from the fact that it had thus stood for ages. I could not but recur also to the fact that once it stood upright; that, although ages had been passed in assuming its present inclination to the earth, the time would probably come when it would actually fall, and the idea would suggest itself with appalling force that that time might be now. The reflection suggested by one of our company that it would be a glorious death, for one thus perishing would be sure of an imperishable name, however pleasing in romantic speculation, had no great power to dispel the shrinking fear produced by the vivid thought of the possibility when on the top of the tower. . . . The *campanile* is not the only leaning tower in Pisa. We observed that several varied from the perpendicular, and the sides of many of the buildings, even parts of the cathedral and the baptistery, inclined at a considerable angle. The soil is evidently unfavorable to the erection of high, heavy buildings."

After a side trip to Leghorn and further loitering along the way, stopping but a short time in Florence, which he purposed to visit and study at his leisure later on, he saw, at nine o'clock on the morning of February 20, the dome of St. Peter's in the distance, and, at two o'clock he and his companions entered Rome through the Porta del Popolo.

Taking lodgings at No. 17 Via de Prefetti, he spent the first few days in a cursory examination of the treasures by which he was surrounded, but he was eager to begin at once the work for which he had received commissions, and on March 7 he writes home: —

"I have begun to copy the 'School of Athens' from Raphael for Mr. R. Donaldson. The original is on the walls of one of the celebrated Camera of Raphael in the Vatican. It is in fresco and occupies one entire side of the room. It is a difficult picture to copy and will occupy five or six weeks certainly. Every moment of my time, from early in the morning until late at night, when not in the Vatican, is occupied in seeing the exhaustless stores of curiosities in art and antiquities with which this wonderful city abounds.

"I find I can endure great fatigue, and my spirits are good, and I feel strong for the pleasant duties of my profession. I feel particularly anxious that every gentleman who has given me a commission shall be more than satisfied that he has received an equivalent for the sum generously advanced to me. But I find that, to accomplish this, I shall need all my strength and time for more than a year to come, and that will be little enough to do myself and them justice. I am delighted with my situation and more than ever convinced of the wisdom of my course in coming to Italy."

Morse's little notebooks and sketch-books are filled with short, abrupt notes on the paintings, religious ceremonies, and other objects of interest by which he is surrounded, but sometimes he goes more into detail. I shall select from these voluminous notes only those which seem to me to be of the greatest interest.

"*March 17*. Mr. Fenimore Cooper and family are here. I have passed many pleasant hours with them, particularly one beautiful moonlight evening visiting the Coliseum. After the Holy Week I shall visit Naples, probably with Mr. Theodore Woolsey, who is now in Rome.

"*March 18.* Ceremonies at the Consistory; delivery of the cardinals' hats. At nine o'clock went to the Vatican; two large fantails with ostrich feathers; ladies penned up; Pope; cardinals kiss his hand in rotation; address in Latin, tinkling, like water gurgling from a bottle. The English cardinal first appeared, went up and was embraced and kissed on each cheek by the Pope; then followed the others in the same manner; then each new cardinal embraced in succession all the other cardinals; after this, beginning with the English cardinal, each went to the Pope, and he, putting on their heads the cardinal's hat, blessed them in the name of the Trinity. They then kissed the ring on his hand and his toe and retired from the throne. The Pope then rose, blessed the assembly by making the sign of the cross three times in the air with his two fingers, and left the room. His dress was a plain mitre of gold tissue, a rich garment of gold and crimson, embroidered, a splendid clasp of gold, about six inches long by four wide, set with precious stones, upon his breast. He is very decrepit, limping or tottering along, has a defect in one eye, and his countenance has an expression of pain, especially as the new cardinals approached his toe.[1]

"The cardinals followed the Pope two and two with their train-bearers. After a few minutes the doors opened again and a procession, headed by singers, entered chanting as they went. The cardinals followed them with their train-bearers; they passed through the Consistory, and thus closed the ceremony of presenting the cardinals' hats.

"A multitude of attendants, in various costumes,

[1] This was Pope Pius VIII.

surrounded the pontiff's throne during the ceremony, among whom was Bishop Dubois of New York. . . .

"Academy of St. Luke's: Raphael's skull; Harlow's picture of the making of a cardinal; said to have been painted in twelve days; I don't believe it. 'The Angels appearing to the Shepherds,' by Bassan — good for color; much trash in the way of portraits. Lower rooms contain the pictures for the premiums; some good; all badly colored. Third Room: Bas-reliefs for the premiums. Fourth Room: Smaller premium pictures; bad. Fifth Room: Drawings; the oldest best, modern bad.

"*Friday, March 19.* We went to St. Peter's to see the procession of cardinals singing in the Capella. Cardinals walked two and two through St. Peter's, knelt on purple velvet cushions before the Capella in prayer, then successively kissed the toe of the bronze image of St. Peter as they walked past it.

"This statue of St. Peter, as a work of art, is as execrable as possible. Part of the toe and foot is worn away and polished, not by the kisses, but by the wiping of the foot after the kisses by the next comer preparatory to kissing it; sometimes with the coat-sleeve by a beggar; with the corner of the cloak by the gentlemen; the shawl by the females; and with a nice cambric handkerchief by the attendant at the ceremony, who wiped the toe after each cardinal's performance. This ceremony is variously performed. Some give it a single kiss and go away; others kiss the toe and then touch the forehead to it and kiss the toe again, repeating the operation three times."

The ceremonies and ritual of the Roman Catholic Church, while appealing to the eye of the artist, were repugnant to his Puritan upbringing, and we find many scornful remarks among his notes. In fact he was, all his life, bitterly opposed to the doctrines of Rome, and in later years, as we shall see, he entered into a heated controversy with a prominent ecclesiastic of that faith in America.

"*March 21.* Chiesa Nuova at seven o'clock in the evening; a sacred opera called 'The Death of Aaron.' Church dark; women not admitted; bell rings and a priest before the altar chants a prayer, after which a boy, about twelve years old apparently, addresses the assembly from the pulpit. I know not the drift of his discourse, but his utterance was like the same gurgling process which I noticed in the orator who addressed the Pope. It was precisely like the fitful tone of the Oneida interpreter.

"*Tuesday, March 23.* At the Vatican all the morning. While preparing my palette a monk, decently habited for a monk, who seemed to have come to the Vatican for the purpose of viewing the pictures, after a little time approached me and, with a very polite bow, offered me a pinch of snuff, which, of course, I took, bowing in return, when he instantly asked me alms. I gave him a *bajocco* for which he seemed very grateful. Truly this is a nation of beggars.

"*Wednesday, March 24.* Vatican all the morning. Saw in returning a great number of priests with a white bag over the left shoulder and begging of the persons they met. This is another instance of begging and robbing confined to one class.

"*Thursday, March 25. Festa* of the Annunciation; Vatican shut. Doors open at eight of the Chiesa di

Minerva; obtained a good place for seeing the ceremony. At half-past nine the cardinals began to assemble; Cardinal Barberini officiated in robes, white embroidered with gold; singing; taking off and putting on mitres, etc.; jumping up and bowing; kissing the ring on the finger of the cardinal; putting incense into censers; monotonous reading, or rather whining, of a few lines of prayer in Latin; flirting censers at each cardinal in succession; cardinals bowing to one another; many attendants at the altar; cardinals embrace one another; after mass a contribution among the cardinals in rich silver plate. Enter the virgins in white, with crowns, two and two, and candles; they kiss the hem of the garment of one of the cardinals; they are accompanied by three officers and exit. Cardinals' dresses exquisitely plaited; sixty-two cardinals in attendance. . . .

"Palazzo Simbaldi: At half-past eight the company began to assemble in the splendid saloon of this palace, to which I was invited. The singers, about forty in number, were upon a stage erected at the end of the room; white drapery hung behind festoons with laurel wreaths (the walls were painted in fresco). Four female statues standing on globes upheld seven long wax-lights; the instrumental musicians, about forty, were arranged at the foot of these statues; *sala* was lighted principally by six glass chandeliers; much female beauty in the room; dresses very various.

"Signora Luigia Tardi sang with much judgment and was received with great applause. A little girl, apparently about twelve years old, played upon the harp in a most exquisite manner, and called forth *bravas* of the Italians and of the foreigners bountifully.

"The manners of the audience were the same as those of fashionable society in our own country, and indeed in any other country; the display in dress, however, less tasteful than I have seen in New York. But, in truth, I have not seen more beauty and taste in any country, combined with cultivation of mind and delicacy of manner, than in our own. At one o'clock in the morning, or half-past six Italian time, the concert was over.

"*Saturday, March 27.* On returning to dinner I found at the post-office, to my great joy, the first letter from America since I left it.

"*Sunday, March 28.* Passion Sunday. Kept awake nearly all last night by a severe toothache; sent for a dentist and had the tooth extracted, for which he had the conscience to ask me three dollars — he took two. Was prevented by this circumstance from going to church this morning; went in the afternoon, and, after church, to St. Peter's; found all the crosses covered with black and all the pictures veiled. There were a great many in the church to hear the music which is considered very fine; some of it I was well pleased with, but it is by no means so impressive as the singing of the nuns at the Trinita di Monti, to which church we repaired at vespers.

"In St. Peter's we found a procession of about forty nuns; some of them were very pretty and their neat white headdresses, and kerchiefs, and hair dressed plain, gave a pleasing simplicity to their countenances. Some looked arch enough and far from serious.

"*Monday, March 29.* Early this morning was introduced to the Chevalier Horace Vernet, principal of the French Academy; found him in the beautiful gardens of the Academy. He came in a *négligé* dress, a cap, or rather turban, of various colors, a parti-colored belt, and a cloak. He received me kindly, walked through the antique gallery of casts, a long room and a splendid collection selected with great judgment.

"*Wednesday, March 31.* Early this morning was waked by the roar of a cannon; learned that it was the anniversary of the present Pope's election. Went to the Vatican; the colonnade was filled with the carriages of the cardinals; that of the new English cardinal, Weld, was the most showy.

"*Thursday, April 1.* Went in the evening to the soirée of the Chevalier Vernet, director of the French Academy. He is a gentleman of elegant manners and sees at his soirées the first society in Rome. His wife is highly accomplished and his daughter is a beautiful girl, full of vivacity, and speaks English fluently. . . . During the evening there was music; his daughter played on the piano and others sang. There was chess, and, at a sideboard, a few played cards. The style was simple, every one at ease like our soirées in America. Several noblemen and dignitaries of the Church were present."

On April 4, Palm Sunday, he attended the services at the Sistine Chapel, which he found rather tedious, with much mummery. Going from there to the cancellerie he describes the following scene: —

"Cardinal Giulio Maria della Somaglia in state on an elevated bed of cloth-of-gold and black embroidered with gold, his head on a black velvet cushion embroidered with gold, dressed in his robes as when alive. He officiated, I was told, on Ash Wednesday. Four wax-lights, two on each side of the bed; great throng of people of all grades through the suite of apartments — the cancellerie — in which he lived; they were very splendid, chiefly of crimson and gold. The cardinal has died unpopular, for he has left nothing to his servants by his will; he directed, however, that no expense should be spared in his funeral, wishing that it might be splendid, but, unfortunately for him, he has died precisely at that season of the year (the Holy Week) when alone it is impossible, according to the church customs, to give him a splendid burial."

"*Wednesday, April 7.* Went to the Piazza Navone, being market-day, in search of prints. The scene here is very amusing; the variety of wares exposed, and the confusion of noises and tongues, and now and then a jackass swelling the chorus with his most exquisite tones.

"At three o'clock went to St. Peter's to see ceremonies at the Sistine Chapel. Cardinals asleep; monotonous bawling, long and tedious; candles put out one by one, fifteen in number; no ceremonies at the altar; cardinals present nineteen in number; seven yawns from the cardinals; tiresome and monotonous beyond description.

"After three hours of this most tiresome chant, all the candles having been extinguished, the celebrated *Miserere* commenced. It is, indeed, sublime, but I think loses much of its effect from the fatigue of body, and mind, too, in which it is heard by the auditors. The *Miserere* is the composition of the celebrated Allegri, and for giving the effect of wailing and lamentation, without injury to harmony, it is one of the most perfect of compositions. The manner of sustaining a strain of concord by new voices, now swelling high, now gradually dying away, now sliding imperceptibly into discord and suddenly breaking into harmony, is admirable. The imagination is alive and fancies thousands of people in the deepest contrition. It closed by the cardinals clapping their hands for the earthquake."

On April 8 (Holy Thursday), Morse went early with Mr. Fenimore Cooper and other Americans to St. Peter's. After describing some of the preliminary ceremonies he continues: —

"Having examined the splendid chair in which he was to be borne, and while he was robing in another apartment, we found that, although we might have a complete view of the Pope and the ceremonies before and after the benediction, yet the principal effect was to be seen below. We therefore left our place at the balcony, where we could see nothing but the crowd, and hastened below. On passing into the hall we were so fortunate as to be just in season for the procession from the Sistine Chapel to the Pauline. The cardinals walked in procession, two and two, and one bore the host, while eight bearers held over him a rich canopy of silver tissue embroidered with gold.

"Thence we hastened to the front of St. Peter's, where, in the centre upon the highest step, we had an excellent view of the balcony, and, turning round, could see the immense crowd which had assembled in the piazza and the splendid square of troops which were drawn up before the steps of the church. Here I had scarcely time to make a hasty sketch, in the broiling sun, of the window and its decorations, before the precursors of the Pope, the two large feather fans, made their appearance on each side of the balcony, which was decorated with crimson and gold, and immediately after the

Pope, with his mitre of gold tissue and his splendid robes of gold and jewels, was borne forward, relieving finely from the deep crimson darkness behind him. He made the usual sign of blessing, with his two fingers raised. A book was then held before him in which he read, with much motion of his head, for a minute. He then rose, extending both his arms — this was the benediction — while at the same moment the soldiers and crowd all knelt; the cannon from the Castle of St. Angelo was discharged, and the bells in all the churches rang a simultaneous peal.

"The effect was exceedingly grand, the most imposing of all the ceremonies I have witnessed. The Pope was then borne back again. Two papers were thrown from the balcony for which there was a great scramble among the crowd."

On Friday, April 9 (Good Friday), many of the ceremonies so familiar to visitors to Rome during Holy Week are described at length in the notebooks, but I shall omit most of these. The following note, however, seems worthy of being recorded: —

"On our way to St. Peter's I ought to have noticed our visit to a palace in which another cardinal (the third who has died within a few days) was lying in state — Cardinal Bertazzoli.

"It is a singular fact, of which I was informed, that about the same time last year three cardinals died, and that it was a common remark that when one died two more soon followed, and the Pope always created three cardinals at a time."

"*Friday, April 16.* At the Vatican all day. I went to the soirée of the Signor Persianis in the evening. Here I had the pleasure of meeting for the first time with the Chevalier Thorwaldsen, the great Danish sculptor, the first now living. He is an old man in appearance having a profusion of grey hair, wildly hanging over his forehead and ears. His face has a strong Northern character, his eyes are light grey, and his complexion sandy; he is a large man of perfectly unassuming manners and of most amiable deportment. Daily receiving homage from all the potentates of Europe, he is still without the least appearance of ostentation. He readily assented to a request to sit for his portrait which I hope soon to take.

"*Tuesday, April 27.* My birthday. How time flies and to how little purpose have I lived!!

"*Wednesday, April 28.* I have noticed a difference in manners between the English, French, and Americans. If you are at the house of a friend and should happen to meet Englishmen who are strangers to you, no introduction takes place unless specially requested. The most perfect indifference is shown towards you by these strangers, quite as much as towards a chair or table. Should you venture a word in the general conversation, they might or might not, as the case may be, take notice of it casually, but coldly and distantly, and even if they should so far relax as to hold a conversation with you through the evening, the moment they rise to go all recognition ceases; they will take leave of every one else, but as soon think of bowing to the chair they had left as to you.

"A Frenchman, on the contrary, respectfully salutes all in the room, friends and strangers alike. He seems to take it for granted that the friends of his friend are at least entitled to respect if not to confidence, and without reserve he freely enters into conversation with you, and, when he goes, he salutes all alike, but no acquaintance ensues.

"An American carries his civility one step further; if he meets you afterwards, in other company, the fact that he has seen you at this friend's and had an agreeable chit-chat is introduction enough, and, unless there is something *peculiar* in your case, he will ever after know you and be your friend. This is not the case with the two former.

"The American is in this, perhaps, too unsuspicious and the others may have good reasons for their mode, but that of the Americans has more of generous sincerity and frankness and kindness in it.

"*Friday, April 30.* Painting all day except two hours at the Colonna Palace — Landi's pictures — horrible!! How I was disappointed. I had heard Landi, the Chevalier Landi, lauded to the skies by the Italians as the greatest modern colorist. He was made a chevalier, elected a member of the Academy at Florence and of the Academy of St. Luke in Rome, and there were his pictures which I was told I must by all means see. They are not merely bad, they are execrable. There is not a redeeming point in a single picture that I saw, not one that would have placed him on a level with the commonest sign-painter in America. His largest work in his rooms at present is the 'Departure of Mary Queen of Scots from Paris.' The story is not told; the figures are not grouped but huddled together; they are not well-drawn individually; the character is vulgar and tame; there is no taste in the disposal of the drapery and ornaments, no effect of *chiaroscuro*. It is flimsy and misty, and, as to color, the quality to which I was specially directed, if total disregard of arrangement, if the scattering of tawdry reds and blues and yellows over the picture, all quarrelling for the precedence; if leather complexions varied by those of chalk, without truth or depth or tone, constitute good color, then are they finely colored. But, if Landi is a colorist, then are Titian and Veronese never more to be admired. In short, I have never met with the works of an artist who had a name like Landi's so utterly destitute of even the shadow of merit. There is but one word which can express their character, they are *execrable!*

"It is astonishing that with such works of the old masters before them as the Italians have, they should not perceive the defects of their own painters in this particular. Cammuccini is the only one among them who possesses genius in the higher departments, and he only in drawing; his color is very bad.

"A funeral procession passed the house to-day. On the bier, exposed as is customary here, was a beautiful young girl, apparently of fifteen, dressed in rich laces and satins embroidered with gold and silver and flowers tastefully arranged, and sprinkled also with real flowers, and at her head was placed a coronet of flowers. She had more the appearance of sleep than of death. No relative appeared near her; the whole seemed to be conducted by the priests and monks and those hideous objects in white hoods, with faces covered except two holes for the eyes."

In early May, Morse, in company with other artists, went on a sketching trip to Tivoli, Subiaco, Vico, and Vara. This must have been one of the happiest periods of his life. He was in Italy, the cradle of the art he loved; he was surrounded by beauty, both natural and that wrought by the hand of man; he had daily intercourse with congenial souls, and home, with its cares and struggles, seemed far away. His notebooks are largely filled with simple descriptions of the places visited, but now and then he indulges in rhapsody. At Subiaco he comes upon this scene: —

"Upon a solitary seat (a fit place for meditation and study), by a gate which shut the part of the terrace near the convent from that which goes round the hill, sat a monk with his book. He seemed no further disturbed by my passing than to give me the usual salutation.

"I stopped at a little distance from him to look around and down into the chasm below. It was enchanting in spite of the atmosphere of the sirocco. The hills covered with woods, at a distance, reminded me of my own country, fresh and variegated; the high peaks beyond were grey from distance, and the sides of the nearer mountains were marked with many a winding track, down one of which a shepherd and his sheep were descending, looking like a moving pathway. No noise disturbed the silence but the distant barking of the shepherd's dog (as he, like a busy marshal, kept the order of his procession unbroken) mixing with the faint murmuring of the waterfall and the song of the birds that inhabited the ilex grove. It was altogether a place suited to meditation, and, were it consistent with those duties which man owes his fellow man, here would be the spot to which one, fond of study and averse to the noise and bustle of the world, would love to retire."

Returning to Rome on June 3, after enjoying to the

full this excursion, from which he brought back many sketches, he found the city given over to ceremony after ceremony connected with the Church. Saint's day followed saint's day, each with its appropriate (or, from the point of view of the New Englander, inappropriate) pageant; or some new church was dedicated and the nights made brilliant with wonderful pyrotechnical displays. He went often with pleasure to the Trinita di Monti, where the beautiful singing of the nuns gave him special pleasure.

Commenting sarcastically on a display of fireworks in honor of St. Francesco Caracciolo, he says: —

"As far as whizzing serpents, wheels, port-fires, rockets, and other varieties of pyrotechnic art could set forth the humility of the saint, it was this night brilliantly displayed."

And again, in describing the procession of the *Corpus Domini*, "the most splendid of all the church ceremonies," it is this which particularly impresses him: —

"Next came monks of the Franciscan and Capuchin orders, with their brown dresses and heads shaved and such a set of human faces I never beheld. They seemed, many of them, like disinterred corpses, for a moment reanimated to go through this ceremony, and then to sink back again into their profound sleep. Pale and haggard and unearthly, the wild eye of the visionary and the stupid stare of the idiot were seen among them, and it needed no stretch of the imagination to find in most the expression of the worst passions of our nature. They chanted as they went, their sepulchral voices echoing through the vaulted piazza, while the bell of St. Peter's, tolling a deep bass drone, seemed a fitting accompaniment for their hymns."

Later, on this same day, while watching a part of the ceremonies on the Corso, he has this rather disagreeable experience: —

"I was standing close to the side of the house when, in an instant, without the slightest notice, my hat was struck off to the distance of several yards by a soldier, or rather a poltroon in a soldier's costume, and this courteous manœuvre was performed with his gun and bayonet, accompanied with curses and taunts and the expression of a demon in his countenance.

"In cases like this there is no redress. The soldier receives his orders to see that all hats are off in this religion of force, and the manner is left to his discretion. If he is a brute, as was the case in this instance, he may strike it off; or, as in some other instances, if the soldier be a gentleman, he may ask to have it taken off. There was no excuse for this outrage on all decency, to which every foreigner is liable and which is not of infrequent occurrence. The blame lies after all, not so much with the pitiful wretch who perpetrates this outrage, as it does with those who gave him such base and indiscriminate orders."

CHAPTER XVII

JUNE 17, 1830 — FEBRUARY 2, 1831

Working hard. — Trip to Genzano. — Lake of Nemi. — Beggars. — Curious festival of flowers at Genzano. — Night on the Campagna. — Heat in Rome. — Illumination of St. Peter's. — St. Peter's Day. — Vaults of the Church. — Feebleness of Pope. — Morse and companions visit Naples, Capri, and Amalfi. — Charms of Amalfi. — Terrible accident. — Flippancy at funerals. — Campo Santo at Naples. — Gruesome conditions. — Ubiquity of beggars. — Convent of St. Martino. — Masterpiece of Spagnoletto. — Returns to Rome. — Paints portrait of Thorwaldsen. — Presented to him in after years by John Taylor Johnston. — Given to King of Denmark. — Reflections on the social evil and the theatre. — Death of the Pope. — An assassination. — The Honorable Mr. Spencer and Catholicism. — Election of Pope Gregory XVI.

DURING all these months Morse was diligently at work in the various galleries, making the copies for which he had received commissions, and the day's record almost invariably begins with "At the Colonna Palace all day"; or, "At the Vatican all day"; or wherever else he may have been working at the time.

The heat of the Roman summer seems not yet to have inconvenienced him, for he does not complain, but simply remarks: "Sun almost vertical, . . . houses and shops shut at noon." He has this to say of an Italian institution: "Lotteries in Rome make for the Government eight thousand scudi per week; common people venture in them; are superstitious and consult *cabaliste* or lucky numbers; these tolerated as they help sell the tickets."

While working hard, he occasionally indulged himself in a holiday, and on June 16 he, in company with three other artists, engaged a carriage for an excursion to Albano, Aricia, and Genzano, "to witness at the latter place the celebrated *festa infiorata*, which occurs every year on the 17th of June."

After spending the night at Albano, which they found crowded with artists of various nationalities and with other sight-seers, "We set out for Genzano, a pleasant walk of a little more than a mile through a winding carriage-road, thickly shaded with fine trees of elm and chestnut and ilex. A little fountain by the wayside delayed us for a moment to sketch it, and we then continued our way through a straight, level, paved road, shaded on each side with trees, into the pretty village of Genzano."

Finding that the principal display was not until the afternoon, they strolled to the Lake of Nemi, "situated in a deep basin, the crater of a volcano." Those Italian lakes which he had so far seen, while lovely and especially interesting from their historical or legendary associations and the picturesque buildings on their shores, seemed to the artist (ever faithful to his native land) less naturally attractive than the lakes with which he was familiar at home — Lake George, Otsego Lake, etc. He had not yet seen Como or Maggiore. Then he touches upon the great drawback to all travelling in Italy: —

"Throughout the day, wherever we went, beggars in every shape annoyed us, nor could we scarcely hear ourselves talk when on the borders of the lake for the swarms which importuned us. A foolish Italian, in the hope probably of getting rid of them, commenced giving a *mezzo biochi* to each, and such a clamor, such devouring eyes, such pushing and bawling, such teasing importunity for more, and from some who had received and concealed their gift, I could not have conceived, nor do I ever wish again to see so disgusting a sight. The foolish fellow who invented this plan of satisfying an Italian beggar's appetite found to his sorrow that, instead of thanks, he obtained curses and an increase of importunity. . . .

"After dinner we again walked to Genzano, whither we found were going great multitudes of every class; elegant equipages and *vetture* racing with each other; donkeys and horses and foot travellers; and not among the least striking were the numbers of women, some of whom were splendidly dressed, all riding on horseback, a foot in each stirrup, and riding with as much ease and fine horsemanship as the men.

"When we arrived at Genzano the decoration of the streets had commenced. Two of the principal and wide streets ascend a little, diverging from each other, from the left side of the common street which goes through the village. The middle of these streets was the principal scene of decoration. On each side of the centre of the street, leaving a good-sized sidewalk, were pillars at a distance of eight or nine feet from each other composed of the evergreen box and tufted at the top with every variety of flowers. They were in many places also connected by festoons of box. The pavement of the street between the pillars in both streets, and for a distance of at least one half a mile, was most exquisitely figured with flowers of various colors, looking like an immense and gorgeously figured carpet.

"The devices were in the following order which I took note of on the spot: first, a temple with four columns of yellow flowers (the flower of the broom) containing an altar on which was the Holy Sacrament. In the pediment of the temple a column surmounted by a half-moon, which is the arms of the Colonna family. Second was a large crown. Third, the Holy Sacrament again with various rich ornaments. Fourth, stars and circles. Fifth, a splendid coat-of-arms as accurate and rich as if emblazoned in permanent colors, with a cardinal's hat and a shield with the words '*prudens*' and '*fidelis*' upon it."

There were twenty of these wonderful floral decorations on the pavement of one street and fourteen on that of the other and all are described in the notes, but I have particularized enough to show their character. The journal continues: —

"All these figures were as elegantly executed as if made for permanency, some with a minuteness truly astonishing. Among other decorations of the day was the free-will offering of one of the people who had it displayed at the side of his shop on a rude pedestal. It was called the 'Flight into Egypt,' and represented Joseph and Mary and the infant on an ass, and all composed of shrubs and flowers. It was, indeed, a most ludicrous-looking affair; Joseph with a face (if such it might be called) of purple flowers and a flaxen wig, dressed in a coarse pilgrim's cape studded over with yellow flowers, was leading by a hay band a green donkey, made of a kind of heath grass, with a tail of lavender and hoofs of cabbage leaves. Of this latter composition were also the sandals of Mary, whose face, as well as that of the *bambino*, was also of purple flowers and shapeless. The frock of the infant was of the gaudiest red poppy. It excited the laughter of almost all who saw it, except now and then some of the ignorant lower classes would touch their hats, cross themselves, and mumble a prayer."

After describing some of the picturesque costumes of the *contadini*, he continues: —

"It was nearly dark before the procession, to which all these preparations had reference, began to move. At length the band of music was heard at the lower end of one of the streets, and a man, in ample robes of scarlet and blue, with a staff, was seen leading the procession, which need be no further described than to say it consisted of the usual quantity of monks chanting, with wax-tapers in their hands, crosses, and heavy, unwieldy banners which endanger the heads of the multitude as they pass; of a fine band of music playing beautiful waltzes and other compositions, and a *quantum suff.* of men dressed in the garb of soldiers to keep the good people uncovered and on their knees.

"The head of the procession had arrived at the top of the street when — crack! pop! — went forty or fifty crackers, which had been placed against the walls of a house near us, and which added wonderfully to the solemnity of the scene, and, accordingly, were repeated every few seconds, forming a fine accompaniment to the waltzes and the chanting of the monks. In a few minutes all the beauty of the flower-carpeted street was trodden out, and the last of the procession had hardly passed before all the flowers disappeared from the pillars, and all was ruin and disorder.

"The procession halted at a temporary altar at the top of the street, and we set out on our return at the same moment down the street, facing the immense multitude which filled the whole street. We had scarcely proceeded a third of the distance down when we suddenly saw all before us uncovered and upon their knees. We alone formed an exception, and we continued our course with various hints from those around us to stop and kneel, which we answered by talking English to each other in a louder tone, and so passed for unchristian *forestieri*, and escaped unmolested, especially as the soldiers were all at the head of the street.

"The effect, however, was exceedingly grand of such a multitude upon their knees, and, could I have divested myself of the thought of the compulsory measures which produced it and the object to which they knelt, the picture of the Virgin, I should have felt the solemnity of a scene which seemed in the outward act to indicate such a universal reverence for Him who alone rightfully claims the homage and devotion of the heart."

Whether this curious custom still persists in Genzano I know not; Baedeker is silent on the subject.

It was nearly dark before they started on the drive back to Rome, and quite dark after they had gone a short distance.

"We passed the tombs of the Horatii and Curiatii, which looked much grander in the light of the torches than in the day, and, driving hastily through Albano, came upon the Campagna once more. It was still more like a desert in the night than in the day, for it was an interminable ocean, and the masses of ruins, coming darker than the rest, seemed like deserted wrecks upon its bosom.

"It is considered dangerous in the summer to sleep while crossing the Campagna; indeed, in certain parts of it, over the Pontine Marshes in July and August, it is said to be certain death, but, if the traveller can keep awake, there is no danger. In spite of the fears which we naturally entertained lest it might be already dangerous, most of us could not avoid sleeping, nor could I, with every effort made for that purpose."

The days following his return to Rome were employed chiefly in copying at the Colonna Palace. The heat was now beginning to grow more oppressive, and we find this note on June 21: —

"In the cool of the morning you see the doors of the cafés thronged with people taking their coffee and sitting on chairs in the streets for some distance round. At *mezzo giorno* the streets are deserted, the shop-doors are closed, and all is still; they have all gone to their *siesta*, their midday sleep. At four o'clock all is bustle again; it seems a fresh morning; the streets and cafés are thronged and the Corso is filled with the equipages of the wealthy, enjoying till quite dark the cool of the evening air.

"The sun is now oppressively warm; the heat is unlike anything I have felt in America. There is a scorching character about it which is indescribable, and the glare of the light is exceedingly painful to the eyes. The evenings are delightful, cool and clear, showing the lustre of the stars gloriously.

"*June 28.* In the evening went to the piazza of St. Peter's to witness the illumination of its magnificent dome and the piazza. The change from the smaller to the larger illumination is one of the grandest spectacles I ever beheld.

"The lanterns which are profusely scattered over it,

showing its whole form in lines of fire, glow brighter and brighter as the evening advances from twilight to dark, till it seems impossible for its brilliance to increase. The crowds below, on foot and in carriages, are in breathless expectation. The great bell of St. Peter's at length strikes the hour of nine, and, at the first stroke, a great ball of light is seen ascending the cross to its pinnacle. This is the signal for thousands of assistants, who are concealed over its vast extent, to light the great lamps, and in an instant all is motion, the whole mass is like a living thing, fire whirling and flashing over it in all directions, till the vast pile blazes as if lighted with a thousand suns. The effect is truly magical, for the agents by whom this change is wrought are invisible."

After the illumination of St. Peter's he went to the Castle of St. Angelo where he witnessed what he describes as the grandest display of fireworks he had ever seen.

"*Tuesday, June 29.* This day is St. Peter's day, the grandest *festa* of the Romish Church. I went with Mr. B. early to St. Peter's to see the ceremonies. The streets were filled with equipages, among which the splendid scarlet-and-gold equipages of the cardinals made the most conspicuous figure. Cardinal Weld's carriage was the richest, and next in magnificence was that of Cardinal Barberini.

"On entering St. Peter's we found it hung throughout with crimson damask and gold and filled with people, except a wide space in the centre with soldiers on each side to keep it open for the procession. We passed up near the statue of St. Peter, who was to-day dressed out in his papal robes, his black face (for it is of bronze) looking rather frightful from beneath the splendid tiara which crowned his head, and the scarlet-and-gold tissue of his robes.

"Having a little time to spare, we followed a portion of the crowd down the steps beside the pedestal of the statue of St. Veronica into the vaults beneath the church, which are illuminated on this festival. Mass was performing in several of the splendid chapels, whose rich decorations of paintings and sculpture are but once a year revealed to the light, save from the obscure glimmering of the wax-taper, which is carried by the guide, to occasional visitors. It is astonishing what a vast amount of expense is here literally buried.

"The ornamented parts are beneath the dome; the other parts are plain, heavy arches and low, almost numberless, and containing the sarcophagi of the Popes and other distinguished characters. The illumination here was confined to a single lamp over each arch, which rather made darkness visible and gave an awful effect to some of the gloomier passages.

"In one part we saw, through a long avenue of arches, an iron-grated door; within was a dim light which just sent its feeble rays upon some objects in its neighborhood, not strong enough to show what they were. It required no great effort of the imagination to fancy an emaciated, spectral figure of a monk poring over a large book which lay before him. It might have been as we imagined; we had not time to examine, for the sound of music far above us summoned us into the regions of day again, and we arrived in the body of the church just as the trumpets were sounding from the balcony within the church over the great door of entrance. The effect of the sound was very grand, reverberating through the lofty arches and aisles of the church.

"We got sight of the head of the procession coming in at the great door, and soon after the Pope, borne in his crimson chair of state, and with the triple crown upon his head and a crimson, gold-embroidered mantilla over his shoulders, was seen entering accompanied by his fan-bearers and other usual attendants, and after him the cardinals and bishops. The Pope, as usual, made the sign of the cross as he went.

"The procession passing up the great aisle went round to the back of the great altar, where was the canopy for the Pope and seats for the cardinals and bishops. The Pope is too feeble to go through the ceremony of high mass; it was, therefore, performed before him by one of the cardinals. There was nothing in this ceremony that was novel or interesting; it was the same monotonous chant from the choir, the same numberless bowings, and genuflections, and puffings of incense, and change of garments, and fussing about the altar. All that was new was the constant bustle about the Pope, kissing of his toe and his hand, helping him to rise and to sit again, bringing and taking away of cushions and robes and tiaras and mitres, and a thousand other little matters that would have enraged any man of weak nerves, if it did not kill him. After two hours of this tedious work (the people in the mean time perfectly inattentive), the ceremony ended, and the Pope was again borne through the church and the crowd returned."

On July 7, Morse, with four friends, left Rome at four o'clock in the morning for Naples, where they arrived on the 11th after the usual experiences; beggars continually marring the peaceful beauty of every scene by their importunities; good inns, with courteous landlords and servants, alternating with wretched taverns and insolent attendants. The little notebook detailing the first ten days' experiences in Naples is missing, and the next one takes up the narrative on July 24, when he and his friends are in Sorrento. I shall not transcribe his impressions of that beautiful town or those of the island of Capri. These places are too familiar to the visitor to Italy and have changed but little in the last eighty years.

From Capri they were rowed over to Amalfi, and narrowly escaped being dashed on the rocks by the sudden rising of a violent gale. At Amalfi they found lodgings in the Franciscan monastery, which is still used as an inn, and here I shall again quote from the journal:—

"The place is in decay and is an excellent specimen of their monastic buildings. It is now in as romantic a state as the most poetic imagination could desire. Here are gloomy halls and dark and decayed rooms; long corridors of chambers, uninhabited except by the lizard and the bat; terraces upon the brow of stupendous precipices; gloomy cells with grated windows, and subterranean apartments and caverns. Remains of rude frescoes stain the crumbling ceiling, and ivy and various wild plants hang down from the opening crevices and cover the tops of the broken walls.

"A rude sundial, without a gnomon, is almost obliterated from the wall of the cloisters, but its motto, '*Dies nostri quasi umbra super terram et nulli est mora,*' still resists the effects of decay, as if to serve the appropriate purpose of the convent's epitaph. At the foot of the long stairs in the great hall is the ruined chapel, its altar broken up and despoiled of its pictures and ornaments.

"We were called to dinner by our host, who was accompanied by his wife, a very pretty woman, two children, the elder carried by the mother, the younger by the old grandparent, an old man of upwards of eighty, who seemed quite pleased with his burden and delighted to show us his charge. The whole family quite prepossessed us in their favor; there seemed to be an unusual degree of affection displayed by the members towards each other which we could not but remark at the time. Our dining apartment was the old *domus refectionis* of the convent, as its name, written over the door which led into the choir, manifested. After an excellent dinner we retired to our chambers for the night.

"*Tuesday, July 27.* We all rested but badly last night. The heat was excessive, the insects, especially mosquitoes, exceedingly troublesome, and the sound of the waves, as they beat against the rocks and chafed the beach in the gusty night, and the howling of the wind, which for a time moaned through the deserted chambers of the convent, all made us restless. I rose several times in the night and, opening my window, looked out on the dark waters of the bay, till the dawn over the mountains warned me that the time for sleep was passing away, and I again threw myself on the bed to rest. But scarcely had I lost myself in sleep before the sound of loud voices below and wailings again waked me. I looked out of my window on the balcony below; it was filled with armed men; soldiers and others like brigands with muskets were in hurried commotion, calling to each other from the balcony and from the terraced steps below.

"While perplexed in conjecturing the meaning of what I saw, Mr. C. called at my door requesting me to rise, as the whole house was in agitation at a terrible accident which had occurred in the night. Dressing in great haste, I went into the contiguous room and, looking out of the window down upon a terrace some thirty feet below, saw the lifeless body of a man, with spots of blood upon his clothes, lying across the font of water. A police officer with a band of men appeared, taking down in writing the particulars for a report. On enquiry I found that the body was that of the old man, the father of our host, whom we had seen the evening before in perfect health. He had the dangerous habit of walking in his sleep and had jumped, it is supposed, in that state out of his chamber window which was directly beneath us; at what time in the night was uncertain. His body must have been beneath me while I was looking from my window in the night.

"Our host, but particularly his brother, seemed for a time almost inconsolable. The lamentations of the latter over the bloody body (as they were laying it out in the room where we had the evening before dined), calling upon his father and mingling his cries with a chant to the Virgin and to the saints, were peculiarly plaintive, and, sounding through the vacant halls of the convent, made a melancholy impression upon us all. . . . Soon after breakfast we went downstairs; several priests and funeral attendants had arrived; the poor old man was laid upon a bed, the room darkened, and four wax-lights burned, two each side of the bed. A short time was taken in preparation, and then upon a bier borne by four bearers, a few preceding it with wax-lights, the body, with the face exposed, as is usual in Italy, was taken down the steep pathway to its long home.

"I could not help remarking the total want of that decent deportment in all those officiating which marks the conduct of those that attend the interment of the dead in our own country. Even the priests seemed to be in high glee, talking and heartily laughing with each other; at what it perplexed me to conjecture.

"I went into the room in which the old man had slept; all was as he had left it. Over the head of the bed were the rude prints of the Virgin and saints, which are so common in all the houses of Italy, and which are supposed to act as charms by these superstitious people. The lamp was on the window ledge where he had placed it, and his scanty wardrobe upon a chair by the bedside. Over the door was a sprig of laurel, placed there since his death.

"The accident of the morning threw a gloom over the whole day; we, however, commenced our sketches from different parts of the convent, and I commenced a picture, a view of Amalfi from the interior of the grotto."

Several of the notebooks are here missing, and from the next in order we find that the travellers must have lingered in or near Sorrento until August 30, when they returned to Naples.

The next entry of interest, while rather gruesome, seems to be worth recording.

"*Wednesday, September 1.* Morning painting. In the afternoon took a ride round the suburbs and visited the Campo Santo. The Campo Santo is the public burial-place. It is a large square enclosure having high walls at the sides and open at the top. It contains three hundred and sixty vaults, one of which is opened every day to receive the dead of that day, and is not again opened until all the others in rotation have been opened.

"As we entered the desolate enclosure the only living beings were three miserable-looking old women gathered together upon the stone of one of the vaults. They sat as if performing some incantation, mumbling their prayers and counting their beads; and one other of the same fraternity, who had been kneeling before a picture, left her position as we entered and knelt upon another of the vaults, where she remained all the time we were present, telling her beads.

"At the farther end of the enclosure was a large portable lever to raise the stones which covered the vaults. Upon the promise of a few *grains* the stone of the vault for the day was raised, and, with the precaution of holding our kerchiefs to our noses, we looked down into the dark vault. Death is sufficiently terrible in itself, and the grave in its best form has enough of horror to make the stoutest heart quail at the thought, but nothing I have seen or read of can equal the Campo Santo for the most loathsome and disgusting mode of burial. The human carcasses of all ages and sexes are here thrown in together to a depth of, perhaps, twenty feet, without coffins, in heaps, most of them perfectly naked, and left to corrupt in a mass, like the offal from a slaughter house. So disgusting a spectacle I never witnessed. There were in sight about twenty bodies, men, women, and children. A child of about six years, with beautiful fair hair, had fallen across the body of a man and lay in the attitude of sleeping.

"But I cannot describe the positions of all without

DE WITT CLINTON

Painted by Morse. Property of the Metropolitan Museum of Art

offence, so I forbear. We were glad to turn away and retrace our steps to our carriage. Never, I believe, in any country, Christian or pagan, is there an instance of such total want of respect for the remains of the dead."

On September 5, he again reverts to the universal plague of beggars in Italy: —

"In passing through the country you may not take notice of a pretty child or seem pleased with it; so soon as you do the mother will instantly importune you for '*qualche cosa*' for the child. Neither can you ask for a cup of cold water at a cottage door, nor ask the way to the next village, nor even make the slightest inquiry of a peasant on any subject, but the result will be '*qualche cosa, signore*.' The first act which a child is taught in Italy is to hold out its hand to beg. Children too young to speak I have seen holding out their little hands for that purpose, and so mechanical is this action that I have seen, in one instance, a boy of nine years nodding in his sleep and yet at regular intervals extending his hand to beg. Begging is here no disgrace; on the contrary, it is made respectable by the customs of the Church."

On September 6, after visiting the catacombs, he goes to the Convent of St. Martino, and indulges in this rhapsody: —

"From a terrace and balcony two views of the beautiful scenery of the city and bay are obtained. From the latter place especially you look down upon the city which is spread like a model far beneath you. There is a great deal of the sublime in thus looking down upon a populous city; one feels for the time separated from the concerns of the world.

"We forget, while we consider the insignificance of that individual man, moving in yonder street and who is scarcely visible to us, that we ourselves are equally insignificant. It is in such a situation that the superiority of the mind over the body is felt. Paradoxical as it may at first seem, its greatness is evinced in the feeling of its own littleness. . . . After gazing here for a while we were shown into the chapel through the choir. . . . In the sacristy is a picture of a dead Christ with the three Marys and Joseph, by Spagnoletto, not only the finest picture by that master, but I am quite inclined to say that it is the finest picture I have yet seen. There is in it a more perfect union of the great qualities of art, — fine conception, just design, admirable disposition of *chiaroscuro*, exquisite color, — whether truth is considered or choice of tone in congruity with the subject's most masterly execution and just character and expression. If any objection were to be made it would, perhaps, be in the particular of character, which, in elevation, in ideality, falls far short of Raphael. In other points it has not its superior."

Returning to Rome on September 14, the only entries I find in the journal for the first few days are, "Painting all day at home," and a short account of a soirée at the Persianis'.

"*Monday, September 20.* Began the portrait of the celebrated sculptor Thorwaldsen. He is a most amiable man and is universally respected. He was never married. In early life he had two children by a mistress; one, a daughter, is now in a convent. It was said that a noble lady of England, of great fortune, became attached to him, and he no less to her, but that the circumstance of his having two illegitimate children prevented a marriage. He is the greatest sculptor of the age. I have studied his works; they are distinguished for simple dignity, just expression, and truth in character and design. The composition is also characterized by simplicity. These qualities combined endow them with that beauty which we so much admire in the works of Greece, whether in literature or art. Thorwaldsen cannot be said to imitate the antique; he rather seems to be one born in the best age of Grecian art; imbued with the spirit of that age, and producing from his own resources kindred works."

The following letter was written by Morse before he left Rome for Naples, but can be more appropriately introduced at this point: —

To the Cavalier Thorwaldsen,

My dear Sir, — I had hoped to have the pleasure of painting your portrait, for which you were so good as to promise to sit, before I left Rome for Naples; but the weather is becoming so oppressive, and there being a party of friends about to travel the same road, I have consented to join them. I shall return to Rome in September or October, and I therefore beg you will allow me then to claim the fulfilment of your kind promise.

What a barrier, my dear sir, is difference of language to social intercourse! I never felt the curse that befell the architects of Babel so sensibly as now, since, as one of the effects of their folly, I am debarred from the gratification and profit which I had promised myself in being known to you.

With highest respect, etc.

Curiously enough, Morse never learned to speak a foreign language fluently, although he could read quite easily French and, I believe, German and Italian, and from certain passages in his journal we infer that he could make himself understood by the Italians.

The portrait of Thorwaldsen was completed and became the property of Philip Hone, Esq., who had given Morse a commission to paint a picture for one hundred dollars, the subject to be left to the discretion of the artist. Mr. Hone valued the portrait highly, and it remained in his gallery until his death. It was then sold and Morse lost track of it for many years. In 1868, being particularly desirous of gaining possession of it again, for a purpose which is explained in a letter quoted a little farther on, he instituted a search for it, and finally learned that it had been purchased by Mr. John Taylor Johnston for four hundred dollars. Before he could enter into negotiations for its purchase, Mr. Johnston heard of his desire to possess it, and of his reasons for this wish, and he generously insisted on presenting it to Morse.

I shall now quote the following extracts from a letter written in Dresden, on January 23, 1868, to Mr. Johnston:—

My dear Sir,—Your letter of the 6th inst. is this moment received, in which I have been startled by your most generous offer presenting me with my portrait of the renowned Thorwaldsen, for which he sat to me in Rome in 1831.

I know not in what terms, my dear sir, to express to you my thanks for this most acceptable gift. I made an excursion to Copenhagen in the summer of 1856, as a sort of devout pilgrimage to the tombs of two renowned Danes, whose labors in their respective departments — the one, Oersted, of science, the other, Thorwaldsen, of art — have so greatly enriched the world.

The personal kindness of the late King Frederick VII, who courteously received me at his castle of Fredericksborg, through the special presentation of Colonel Raslof (more recently the Danish Minister at Washington); the hospitalities of many of the principal citizens of Copenhagen; the visits to the tomb and museum of the works of Thorwaldsen, and to the room in which the immortal Oersted made his brilliant electro-magnetic discovery; the casual and accidental introduction and interview with a daughter of Oersted, — all created a train of reflection which prompted me to devise some suitable mode of showing to these hospitable people my appreciation of their friendly attentions, and I proposed to myself the presentation to His Majesty the King of Denmark of this portrait of Thorwaldsen, for which he sat to me in Rome, and with which I knew he was specially pleased.

My desire to accomplish this purpose was further strengthened by the additional attention of the King at a later period in sending me the decoration of his order of the Danebrog. From the moment this purpose was formed, twelve years ago, I have been desirous of obtaining this portrait, and watching for the opportunity of possessing it again.

Here follows a detailed account of the circumstances of the painting of the portrait and of its disappearance, with which we are familiar, and he closes by saying: —

"This brief history will show you, my dear sir, what a boon you have conferred upon me. Indeed, it seems like a dream, and if my most cordial thanks, not merely for the *gift*, but for the graceful and generous manner in which it has been offered, is any compensation, you may be sure they are yours.

"These are no conventional words, but they come from a heart that can gratefully appreciate the noble sentiments which have prompted your generous act."

Returning from this little excursion into later years, I shall take up the narrative again as revealed in the notebooks. While occasionally visiting the opera and the theatre, Morse does not altogether approve of them, and, on September 21, he indulges in the following reflections on them and on the social evil: —

"No females of openly dissolute character were seen, such as occupy particular parts of the theatre in England and America. Indeed, they never appear on the streets of Rome in that unblushing manner as in London, and even in New York and Philadelphia. It must not from hence be inferred that vice is less frequent here than elsewhere; there is enough of it, but it is carried on in secret; it is deeper and preys more on the vitals of society than with us. This vice with us, like a humor on the skin, deforms the surface, but here it infects the very heart; the whole system is affected; it is rotten to the core.

"Theatres here and with us are different institutions. Here, where thousands for want of thought, or rather matter for thought, would die of ennui, where it is an object to escape from home and even from one's self, the theatre serves the purpose of a momentary excitement. A new piece, a new performer, furnishes matter for conversation and turns off the mind from the discussion of points of theology or politics. The theatre is therefore encouraged by the Government and is guarded against the abuses of popular assemblage by strong military guards.

"But what have we to do with theatres in America? Have we not the whole world of topics for discussion or conversation open to us? Is not truth in religion, politics, and science suffered to be assailed by enemies freely, and does it not, therefore, require the time of all intelligent men to study, and understand, and defend, and fortify themselves in truth? Have we time to throw away?

"More than this, have we not homes where domestic endearments charm us, where domestic duties require our attention, where the relations of wife, of husband, of children have the ties of mutual affection and mutual confidence to attach us to our firesides? Need we go abroad for amusement? Can the theatre, with all its tinsel finery, attract away from home the man who has once tasted the bliss of a happy family circle? Is there no pleasure in seeing that romping group of children, in the heyday of youth, amuse themselves ere they go to rest; is there no pleasure in studying the characters of your little family as they thus undisguisedly display themselves, and so give you the opportunity of directing their minds to the best advantage? Is there no amusement in watching the development of the infant mind and in assisting its feeble efforts?

"He must be of most unsocial mould who can leave the thousand charms of home to pass those precious hours in the noxious atmosphere of a theatre, there to be excited, to return at midnight, to rise from a late bed, to pass the best hours of the day in a feverish reverie succeeded by the natural depression which is sure to follow, and to crave a renewed indulgence. Repeated renewal causes indifference and ennui to succeed, till excitement is no longer produced, but gives place to a habit of listless indifference, or a spirit of captious criticism.

"*Monday, November 8, 1830.* A year to-day since I left home.

"*Tuesday, November 9.* Ignorance at post-office. Sent letters for United States to England, because the United States belong to England!

"*Wednesday, December 1.* Many reports for some days past prepared us for the announcement of the death of the Pope, Pius VIII, who died last evening at nine o'clock at the Quirinal Palace."

The ceremonies connected with the funeral of the dead Pope and with the choice of his successor are described at great length, and the eye of the artist was fascinated by the wealth of color and the pomp, while his Protestant soul was wearied and disgusted by the tediousness and mummery of the ceremonials.

"*December 14.* Much excitement has been created by fear of revolution, but from what cause I cannot learn. Many arrests and banishments have occurred, among whom are some of the Bonaparte family. Artists are suspected of being Liberals.

"An assassination occurred at one of the altars in St. John Lateran a few weeks ago. A young man, jealous of a girl, whom he thought to be more partial to another, stabbed her to the heart while at mass.

"*Saturday, January 1, 1831.* At the beginning of the year, as with us, you hear the salutation of '*felicissimo capo d'anno,*' and the custom of calling and felicitating friends is nearly the same as in New York, with this difference, indeed, that there is no cheer in Rome as with our good people at home.

"*Friday, January 14.* In the afternoon Count Grice and the Honorable Mr. Spencer, son of Earl Spencer, who has within a few years been converted to the Catholic faith, called. Had an interesting conversation with him on religious topics, in which the differences of the Protestant and Catholic faiths were discussed; found him a candid, fair-minded man, but evidently led away by a too easy assent to the sophistry and fable which have been dealt out to him. He gave me a slight history of his change; I shall see him again.

"*Tuesday, January 18.* Called with Count Grice on the Honorable Mr. Spencer at the English College and was introduced to the rector, Dr. Wiseman. After a few moments went into the library with Mr. Spencer and commenced the argument, in which being interrupted we retired to his room, where for three hours we discussed various points of difference in our faith. Many things I urged were not answered, such as the fruits of the Catholic religion in the various countries where it prevails; the objection concerning forbidding to marry; idolatry of the Virgin Mary, etc., etc.; yet there is a gentleness, an amiability in the man which makes me think him sincere but deceived.

"*Wednesday, February 2.* Went this morning at ten o'clock to hear a sermon by Mr. Spencer in the chapel of the English College. It was on the occasion of the *festa* of the purification of the Virgin. Many parts were good, and I could agree with him in the general scope of his discourse.

"While we were in the chapel the cannon of St. Angelo announced the election of the new Pope. I hurried to the Quirinal Palace to see the ceremony of announcing him to the people, but was too late. The ceremony was over, the walled window was broken down and the cardinals had presented the new Pope on the balcony. He is Cardinal Cappellari who has taken the title of Gregorio XVI. To-morrow he will go to St. Peter's."

CHAPTER XVIII

FEBRUARY 10, 1831 — SEPTEMBER 12, 1831.

Historic events witnessed by Morse. — Rumors of revolution. — Danger to foreigners. — Coronation of the new Pope. — Pleasant experience. — Cause of the revolution a mystery. — Bloody plot foiled. — Plans to leave for Florence. — Sends casts, etc., to National Academy of Design. — Leaves Rome.— Dangers of the journey.— Florence.— Description of meeting with Prince Radziwill in Coliseum at Rome. — Copies portraits of Rubens and Titian in Florence. — Leaves Florence for Venice. — Disagreeable voyage on the Po. — Venice, beautiful but smelly. — Copies Tintoret's "Miracle of the Slave." — Thunderstorms. — Reflections on the Fourth of July. — Leaves Venice. — Recoaro. — Milan. — Reflections on Catholicism and art. — Como and Maggiore. — The Rigi. — Schaffhausen and Heidelberg. — Evades the quarantine on French border. — Thrilling experience. — Paris.

It was Morse's good fortune to have been a spectator, at various times and in different places, of events of more or less historical moment. We have seen that he was in England during the War of 1812; that he witnessed the execution of the assassin of a Prime Minister; that he was a keen and interested observer of the festivities in honor of a Czar of Russia, a King of France, and a famous general (Blücher); and although not mentioned in his correspondence, he was fond of telling how he had seen the ship sailing away to distant St. Helena bearing the conquered Napoleon Bonaparte into captivity. Now, while he was diligently pursuing his art in Rome, he was privileged to witness the funeral obsequies of one Pope and the ceremonies attendant upon the installation of his successor. In future years the same good fortune followed him.

His presence on these occasions was not always unattended by danger to himself. His discretion during the years of war between England and America saved him from possible annoyance or worse, and now again in Rome he was called upon to exercise the same virtue, for the Church had entered upon troublous times, and soon the lives of foreigners were in danger, and many of them left the city.

On Thursday, February 10, there is this entry in the journal: "The revolutions in the Papal States to the north at Bologna and Ancona, and in the Duchy of Modena, have been made known at Rome. Great consternation prevails." We learn further that, on February 12, "Rumors of conspiracy are numerous. The time, the places of rendezvous, and even the numbers are openly talked of. The streets are filled with the people who gaze at each other inquisitively, and apprehension seems marked on every face. The shops are shutting, troops are stationed in the piazzas, and everything wears a gloomy aspect. At half-past seven a discharge of musketry is heard. Among the reports of the day is one that the Trasteverini have plotted to massacre the *forestieri* in case of a revolt."

While the festivities of the Carnival were, on account of these disturbances, ordered by the Pope to be discontinued, the religious ceremonies were still observed, and, going to St. Peter's one day — "to witness the ceremonies of consecration as a bishop and coronation as a king of the Pope" — Morse had this pleasant experience: —

"The immense area seemed already filled; a double line of soldiers enclosed a wide space, from the great door through the middle of the church, on each side of the altar, and around the richly enclosed space where were erected the two papal thrones and the seats for the cardinals. Into this soldier-invested space none but the privileged were permitted to enter; ambassadors, princes, dukes, and nobles of every degree were seen, in all their splendor of costume, promenading.

"I was with the crowd without, making up my mind to see nothing of the ceremonies, but, being in full dress, and remembering that, on former occasions, I had been admitted as a stranger within the space, I determined to make the effort again. I therefore edged myself through the mass of people until I reached the line of soldiers, and, catching the eye of the commanding officer as he passed by, I beckoned to him, and, as he came to me, I said, '*Sono un Americano, un forestiero, signore,*' which I had no sooner said than, taking me by the hand, he drew me in, and, politely bowing, gave me leave to go where I pleased."

From this point of vantage he had an excellent view of all the ceremonies, which were much like the others he had witnessed and do not need to be described.

He wanted very much to go to Florence at this time to fulfil some of the commissions he had received for copies of famous paintings in that city, but his departure was delayed, for, as he notes on February 13: —

"There are many alarming rumors, one in particular that the Trasteverini and Galleotti, or galley slaves, have been secretly armed by the Government, and that the former are particularly incensed against the *forestieri* as the supposed instigators of the revolution. . . . These facts have thrown us all into alarm, for we know not what excesses such men may be guilty of when excited by religious enthusiasm to revenge themselves on those they call heretics. We are compelled, too, to remain in

Rome from the state of the country, it being not safe to travel on account of brigands who now infest the roads.

"*February 15.* I have never been in a place where it was so difficult to ascertain the truth as in this city. I have enquired the reason of this movement hostile to the Government, but cannot ascertain precisely its object. Some say it is to deprive the Pope of his temporal power, — and some Catholics seem to think that their religion would flourish the better for it; others that it is a plan, long digested, for bringing all Italy under one government, having it divided into so many federative states, like the United States. . . .

"The Trasteverini seem to be a peculiar class, proud, as believing themselves to be the only true descendants from the ancient Romans, and, therefore, hating the other Romans. Poor from that very pride; ignorant and attached to their faith, they are the class of all others to be dreaded in a season of anarchy. It is easy by flattery, by a little distribution of money, and by a cry of danger to their religion, to rouse them to any degree of enthusiasm, and no one can set bounds to the excesses of such a set of fiends when let loose upon society.

"The Government at present have them in their interest, and, while that is the case, no danger is to be dreaded. It is in that state of anarchy which, for a longer or shorter period, intervenes in the changes of government, between the established rule of the one and of the other, that such a class of men is to be feared.

"*February 17.* The plan said to have been determined on by the conspirators was this: The last night of the Carnival was fixed for the execution of the plan. This was Tuesday night when it is customary to have the *moccoletti*, or small wax-candles, lighted by the crowd. The conspirators were each to be placed, as it were by accident, by the side of a soldier (which in so great a crowd could be done without suspicion), and, when the cannon fired which gave the signal for closing the course, it was also to serve as a signal for each one to turn upon the soldier and, by killing him, to seize his arms. This would, indeed, have been a bloody scene, and for humanity's sake it is well that it was discovered and prevented.

"*February 20.* I learn that the Pope is desirous of yielding to the spirit of the times, and is disposed to grant a constitution to the people, but that the cardinals oppose it. He is said also to be prepared to fly from Rome, and even has declared his intention of resigning the dignity of Pope and retiring again to the solitude of the convent.

"*February 24.* It seems to be no longer doubtful that a revolutionary army is approaching Rome from the revolted provinces, and that they advance rapidly. . . . The city is tranquil enough; no troops are seen, except at night a sentinel at some corner cries as you pass, '*Chi viva?*' and you are obliged to cry, '*Il Papa*'; which one may surely do with a good conscience, for he is entitled to great respect for his personal character.

"*February 25.* Went to-day to get my passport viséed for Florence, whither I intended to go on Tuesday next, but am advised by the consul and others not to risk the journey at present, as it is unsafe."

I break the continuity of the narrative for a moment to note that while Morse was making copies of famous paintings in Rome, and studying intelligently the works of the old masters, he was not forgetful of the young academy at home, which he had helped to found and of which he was still president. On March 1 he writes jubilantly to the secretary, J. L. Morton, that he has succeeded in obtaining by gift a number of casts of ancient and modern sculpture which he will send home by the first opportunity. Among the generous donors he mentions Thorwaldsen, Daniel Coit, Esq., Richard Wyatt, Esq., Signor Trentanove, and George Washington Lee, Esq. He adds at the end of the letter: —

"I leave Rome immediately and know not when I shall be allowed to rest, the revolution here having turned everything into confusion, rendering the movements of travellers uncertain and unsafe, and embarrassing my studies and those of other artists exceedingly. I shall try to go to Florence, but must pass through the two hostile armies and through a country which, in a season of confusion like the present, is sure to be infested with brigands. If I reach Florence in safety and am allowed to remain, which is somewhat doubtful, you shall hear of me again, either directly or through my brothers."

Mr. Morton, answering this letter on May 22, informs Morse of his reëlection as president of the National Academy of Design, and adds: "By the by, talking of coming back, do try and make your arrangements as soon as possible. We want you very much, if it is only to set us all right again. We begin to feel the want of our *Head Man*."

Reverting to the journal again, we find this note: "*March 3.* For some days past I have been engaged in packing up and taking leave, and yesterday was introduced by the Count le Grice to Cardinal Weld, who received me very politely, presented me with a book, and sent me two letters of introduction to London."

On March 4, Morse, with four companions, started from Rome on the seemingly perilous journey to Florence. They passed through the lines of both armies, but, contrary to their expectations, they were most courteously treated by the officers on both sides. It is true that they learned afterwards that they came near being arrested at Civita Castellana, where the Papal army was assembled in force, for — "When we took leave of the Marquis at Terni he told us that it was well we left Civita Castellana as we did, for an order for our arrest was making out, and in a few minutes more we should not have been allowed to leave the place. Indeed, when I think of the case, it was a surprising thing that we were allowed to go into all parts of the place, to see their position, to count their men and know their strength, and then to immediately pass over to their enemy and to give him, if we chose, all the information that any spy could have given."

It is not within the province of this work to deal at length with the political movements of the times. As we have seen, Morse was fortunate in avoiding danger, and we learn from history that this revolt, which threatened at one time to become very serious, was eventually suppressed by the Papal arms aided by the Austrians.

Having passed safely through the zone of danger, they travelled on, and, on March 9: —

"At half-past three the *beautiful city* was seen to our left reposing in sunshine in the wide vale of the Arno. The Duomo and the Campanile were the most conspicuous objects. At half-past four we entered Florence and obtained rooms at the Leone Bianco in the Via Vigna Nuova.

"*March 10.* We found to-day, to our great discomfiture, that we are allowed by the police to stay but three days in the city. No entreaties through our consul, nor offers of guaranty on his part, availed to soften towards us the rigor of the decree, which they say applies to all foreigners. I have written to our consul at Leghorn to petition the Government for our stay, as Mr. Ombrosi, the United States Consul here, is not accredited by the Government."

He must have succeeded in obtaining permission to remain, although the fact is not noted in the journal, for the next entry is on April 11, and finds him still in Florence. It begins: "Various engagements preventing my entering regularly in my journal every day's events as they occurred, I have been compelled to make a gap, which I fill up from recollection."

Before following him further, however, I shall quote from a letter written to his brothers on April 15, but referring to events which happened some time before: —

"We have recently heard of the disasters of the Poles. What noble people; how deserving of their freedom. I must tell you of an interesting circumstance that occurred to me in relation to Poland. It was in the latter part of June of last year, just as I was completing my arrangements for my journey to Naples, that I was tempted by one of those splendid moonlight evenings, so common in Italy, to visit once more the ruins of the Coliseum. I had frequently been to the Coliseum in company, but now I had the curiosity to go alone — I wished to enjoy, if possible, its solitude and its solemn grandeur unannoyed by the presence of any one.

"It was eleven o'clock when I left my lodgings and no one was walking at that hour in the solitary streets of Rome. From the Corso to the Forum all was as still as in a deserted city. The ruins of the Forum, the temples and pillars, the Arch of Titus and the gigantic arcade of the Temple of Peace, seemed to sleep in the gravelike stillness of the air. The only sound that reached my ears was that of my own footsteps. I slowly proceeded, stopping occasionally, and listening and enjoying the profound repose and the solemn, pure light, so suited to the ruined magnificence around me. As I approached the Coliseum the shriek of an owl and the answering echo broke the stillness for a moment, and all was still again.

"I reached the entrance, before which paced a lonely sentinel, his arms flashing in the moonbeams. He abruptly stopped me and told me I could not enter. I asked him why. He replied that his orders were to let no one pass. I told him I knew better, that he had no such orders, that he was placed there to protect visitors, and not to prevent their entrance, and that I should pass. Finding me resolute (for I knew by experience his motive was merely to extort money), he softened in his tone, and wished me to wait until he could speak to the sergeant of the guard. To this I assented, and, while he was gone, a party of gentlemen approached also to the entrance. One of them, having heard the discourse between the sentinel and myself, addressed me. Perceiving that he was a foreigner, I asked him if he spoke English. He replied with a slight accent, 'Yes, a little. You are an Englishman, sir?' 'No,' I replied, 'I am an

American from the United States.' 'Indeed,' said he, 'that is much better'; and, extending his hand, he shook me cordially by the hand, adding, 'I have a great respect for your country and I know many of your countrymen.' He then mentioned Dr. Jarvis and Mr. Cooper, the novelist, the latter of whom he said was held in the greatest estimation in Europe, and nowhere more so than in his country, Poland, where his works were more sought after than those of Scott, and his mind was esteemed of an equal if not of a superior cast.

"This casual introduction of literary topics furnished us with ample matter for conversation while we were not engaged in contemplating the sublime ruins over which, when the sentinel returned, we climbed. I asked him respecting the literature of Poland, and particularly if there were now any living poets of eminence. He observed: 'Yes, sir, I am happily travelling in company with the most celebrated of our poets, Meinenvitch'; and who, as I understood him, was one of the party walking in another part of the ruins.

"Engaged in conversation we left the Coliseum together and slowly proceeded into the city. I told him of the deep interest with which Poland was regarded in the United States, and that her heroes were spoken of with the same veneration as our own. As some evidence of this estimation I informed him of the monument erected by the cadets of West Point to the memory of Kosciusko. With this intelligence he was evidently much affected; he took my hand and exclaimed with great enthusiasm and emphatically: 'We, too, sir, shall be free; the time is coming; we too shall be free; my unhappy country will be free.' (This was before the revolution in France.)

"As I came to the street where we were to part he took out his notebook, and, going under the lamp of a Madonna, near the Piazza Colonna, he wished me to write my name for him among the other names of Americans which he had treasured in his book. I complied with his request. In bidding me adieu he said: 'It will be one of my happiest recollections of Rome that the last night which I passed in this city was passed in the Coliseum, and with an American, a citizen of a free country. If you should ever visit Warsaw, pray enquire for Prince ——; I shall be exceedingly glad to see you.'

"Thus I parted with this interesting Pole. That I should have forgotten a Polish name, pronounced but once, you will not think extraordinary. The sequel remains to be told. When the Polish revolution broke out, what was my surprise to find the poet Meinenvitch and a prince, whose name seemed like that which he pronounced to me, and to which was added — 'just returned from Italy' — among the first members of the provisional government."

Morse assured himself afterwards, and so noted it in his journal, that this chance acquaintance was Prince Michael Jerome Radziwill, who had served as lieutenant in the war of independence under Kosciusko; fought under Napoleon in Russia (by whom he was made a brigadier-general); and, shortly after the meeting in the Coliseum, was made general-in-chief of the Polish army. After the defeat of this army he was banished to central Russia until 1836, when he retired to Dresden.

Reverting again to the notebooks, we find that Florence, with her wealth of beauty in architecture, sculpture, and painting, appealed strongly to the artist, and the notes are chiefly descriptions of what he sees, and which it will not be necessary to transcribe. He had, during all the time he was in Italy, been completing, one after another, the copies for which he had received commissions, and had been sending them home. He thus describes to his friend, Mr. Van Schaick, the paintings made for him: —

"*Florence, May 12, 1831.* I have at length completed the two pictures which you were so kind as to commission me to execute for you, and they are packed in a case ready to send to you from Leghorn by the first opportunity, through Messrs. Bell, de Yongh & Co. of that city.

"As your request was that these pictures should be heads, I have chosen two of the most celebrated in the gallery of portraits in the Florence Gallery. These are the heads of Rubens and Titian from the portraits by themselves. As the portraits of the two great masters of color they will alone be interesting, but they are more so from giving a fair specimen of their two opposite styles of color. That of Rubens, from its gaiety, will doubtless be more popular, but that of Titian, from its sobriety and dignity, pleases me better. In hanging the pictures they should be placed apart. The styles are so opposed that, were they placed near to each other, they would mutually affect each other unfavorably. Rubens may be placed in more obscurity, but Titian demands to be more in the light.

"I have no time to add, as I am preparing to leave Florence on Monday for Bologna and Venice."

Travelling in Italy in those days was fraught with many annoyances, for, in addition to the slow progress made in the *vetture*, there seems to have been (judging from the journal) a *dogana*, or custom-house, every few miles, where the luggage and clothing of travellers were examined, sometimes hastily and courteously, sometimes with more rigor. And yet this leisurely rate of progress, the travellers walking up most of the hills, must have had a charm unknown to the present-day tourist, who is whisked unseeing through the most characteristic parts of a foreign country. The beautiful scenery of the Apennines was in this way enjoyed to the full by the artist, but I shall not linger over the journey nor shall I include any notes concerning Bologna. He found the city most interesting — "A piece of porphyry set in verd antique" — and those to whom he had letters of introduction more hospitable than in any other city in Italy.

From Bologna the route lay through Ferrara and then to Pontelagoscuro on the river Po, where he was to take the courier boat for Venice, down the Po and through a canal. To add to the discomforts of this part of the trip it rained steadily for several days, and, on May 22, Morse paints this dreary picture: —

"When we waked this morning we found it still raining and, apparently, so to continue all day. The rainy day at a country inn, so exquisitely described by Irving in all its disagreeable features, is now before us. A solitary inn with nothing indoors to attract; cold and damp and dark. The prospect from the windows is a low muddy foreground, the north bank of the muddy Po; a pile of brushwood, a heap of offal, a melancholy group of cattle, who show no other signs of life than the occasional sly attack by one of them upon a poor, dripping, half-starved dog, who, with tail between his legs, now and then ventures near them to search for his miserable meal. Beyond, on the river, a few barks silently lying upon the stream, and on the opposite bank some buildings with a church and a campanile dimly seen through the mist. After coffee we were obliged to go to the *dogana* to see to the searching of all our trunks and luggage. The principals were present and we were not severely searched. A Frenchman, however, who had come on a little before us, was stripped to his skin, some papers were found upon him, and I understand he has made his escape and they are now searching for him. . . .

"At 2.30, after having dined, we waded through the mud in a pelting rain to the *dogana* for our luggage, and, after getting completely wet, we embarked on board the courier boat, with a cabin seven feet long, six feet wide, and six high, into which six of us, having a gentleman from Trieste and his mother added to our number, were crowded, with no beds. . . . Rain, rain, rain! ! ! in torrents, cold and dreary through a perfectly flat country. . . . At ten o'clock we arrived at a place called Cavanella, where is a *locanda* upon the canal which should have been open to receive us, but they were all asleep and no calling would rouse them. So we were obliged to go supperless to bed, and such a bed! There being no room to spread mattresses for six in the cabin, three dirty mattresses, without sheets or blankets, were laid on the floor of the forward cabin (if it might so be called). This cabin was a hole down into which two or three steps led. We could not stand upright, — indeed, kneeling, our heads touched the top, — and when stretched at full length the tallest of us could touch with his head and feet from side to side. But, it being dreary and damp without and we being sleepy, we considered not the place, nor its inconveniences, nor its little pests which annoyed us all night, nor its vicinity to a magazine of cheese, with which the boat was laden and the odors from which assailed us. We lay down in our clothes and slept; the rain pattering above our heads only causing us to sleep the sounder."

Continuing their leisurely journey in this primitive manner, the rain finally ceasing, but the sky remaining overcast and the weather cold and wintry, they reached Chioggia, and "At 11.30, the towers and spires of Venice were seen at a distance before us rising from the sea." Venice, of course, was a delight to Morse's eye, but his nose was affected quite differently, for he says: "Those that have resided in Venice a long time say it is not an unhealthy place. I cannot believe it, for the odors from the canals cannot but produce illness of some kind. That which is constantly offensive to any of our organs of sense must affect them injuriously."

Several severe thunderstorms broke over the city while he was there, and one was said to be the worst which had been known within the memory of the oldest inhabitant. After describing it he adds: "I was at the Academy. The rain penetrated through the ceiling at the corner of the picture I was copying — 'The Miracle of the Slave,' by Tintoret — and threatened injury to it, but happily it escaped."

On June 19, he thus moralizes: "The Piazza of St. Mark is the great place of resort, and on every evening, but especially on Sundays or *festas*, the arcades and cafés are crowded with elegantly dressed females and their gallants. Chairs are placed in great numbers under the awnings before the cafés. A people that have no homes, who are deprived from policy of that domestic and social intercourse which we enjoy, must have recourse to this empty, heartless enjoyment; an indolent enjoyment, when all their intercourse, too, is in public, surrounded by police agents and soldiers to prevent excess. Hallam, in his 'Middle Ages,' has this just reflection on the condition of this same city when under the Council of Ten: 'But how much more honorable are the wildest excesses of faction than the stillness and moral degradation of servitude.' Quiet is, indeed, obtained here, but at what immense expense! Expense of wealth, although excessive, is nothing compared with the expense of morality and of all intellectual exercise."

On June 23, he witnessed another thunderstorm from the Piazza of St. Mark: —

"The lightning, flashing in the dark clouds that were gathering from the Tyrolese Alps, portended another storm which soon burst over us and hastened the conclusion of the music. The lightning was incessant. I stood at the corner of the piazza and watched the splendid effects of lights and darks, in a moment coming and in a moment gone, on the campanile and church of St. Mark's. It was most sublime. The gilt statue of the angel on the top of the campanile never looked so sublime, seeming to be enveloped in the glory of the vivid light, and, as the electric fluid flashed behind it from cloud to cloud incessantly, it seemed to go and come at the bidding of the angel."

This sounds almost like a prophetic vision, written by the pencil of the man who, in a few years from then, was to make the lightning go and come at his bidding.

"*July 4.* This anniversary of the day of our national birth found but two Americans in Venice. We met in the evening over a cup of coffee and thought and talked of the happiest of countries. We had no patriotic toasts, but the sentiments of our hearts were — 'Peace be within thy walls and prosperity within thy palaces.' Never on any anniversary of our Independence have I felt so strongly the great reason I have for gratitude in having been born in such a country. When I think of the innumerable blessings we enjoy over every other country in the world, I am constrained to praise God who hath made us to differ, for 'He hath not dealt so with any nation, and as for his judgments, we have not known them.' While pestilence and famine and war surround me here in these devoted countries, I fix my thoughts on one bright spot on earth; truly (if our too ungrateful countrymen would but see it), truly a terrestrial paradise."

This attack of nostalgia was probably largely due to atmospheric conditions, for at least one thunderstorm seems to have been a matter of daily occurrence. This, added to the noisome odors arising from the canals, affected his health, for he complains of feeling more unwell than at any time since he left home. It must, therefore, have been with no feelings of great regret that he packed his belongings and prepared to leave Venice with a companion, Mr. Ferguson, of Natchez, on the 18th of July. His objective point was Paris, but he planned to linger by the way and take a leisurely course through the

Italian lake region, Switzerland, and Germany. The notebooks give a detailed but rather dry account of the daily happenings. It was, presumably, Morse's intention to elaborate these, at some future day, into a more entertaining record of his wanderings; but this was never done. I shall, therefore, pass on rapidly, touching but lightly on the incidents of the journey, which were, in the main, without special interest. The route lay through Padua, Vicenza, Verona, and Brescia to Milan. From Vicenza a side trip was made to the watering-place of Recoaro, where a few days were most delightfully spent in the company of the English consul at Venice, Mr. Money, and his family.

"Recoaro, like all watering-places, is beginning to be the resort of the fashionable world. The Grand Duchess of Tuscany is now here, and on Saturday the Vice-Queen of Italy is expected from Milan to visit her aunt, the Grand Duchess. . . . Towards evening parties of ladies and gentlemen are seen promenading or riding on donkeys along the brows of the mountains and among the trees, and many priests are seen disfiguring the landscape with their tasteless, uncouth dresses; most of them coming, I was informed on the best authority, for the purpose of gambling and dissipating that time of which, from the trifling nature of their duties and the almost countless increase of their numbers, they have so much to spare. Cards have the most fascination for them."

Another incident of the stay at Recoaro is worth recording. Referring to the family of Mr. Money, he says: —

"In the afternoon took an excursion on donkeys with

the whole family among the wild and romantic scenery. In returning, while riding by the side of Mr. Money and in conversation with him, my donkey stumbled upon his knees and threw me over his head, without injury to me, but Mrs. Money, who was just before me, seeing the accident, was near fainting and, during the rest of the day, was invisible. I was somewhat surprised at the effect produced on her until I learned that the news of the loss of her son in India by a fall from his horse, which had recently reached her, had rendered her nerves peculiarly sensitive."

Two days later, however, he joined them in another excursion.

"On returning we stopped to take tea at Mrs. Ireland's lodgings, an English lady who is here with her two daughters, accomplished and highly agreeable people. I was told by them that after I left Rome a most diabolical attempt was made to poison the English artists who had made a party to Grotto Ferrata. They were mistaken by the persons who attempted the deed for Germans. They all became exceedingly ill immediately after dinner, and, as the wine was the only thing they had taken there, having brought their food with them, it was suspected and a strong solution of copper was proved to be in it. I was told that Messrs. Gibson and Desoulavy suffered a great deal, the latter being confined to his bed for three weeks. Had I been in Rome it is more than probable I should have been of their party, for I had never visited Grotto Ferrata, and the company of those with whom I had associated would have induced me to join them without a doubt."

Morse enjoyed his stay at Recoaro so much that he

was persuaded by his hospitable friends to prolong his visit for a few days longer than he had planned, but, on July 27, he and his friend Mr. Ferguson bade adieu and proceeded on their journey. Verona and Brescia were visited and on July 29 they came to Milan. The cathedral he finds "a most gorgeous building, far exceeding my conception of it"; and of the beautiful street of the Corso Porta Orientale he says: "It is wider than Broadway and as superior as white marble palaces are to red brick houses. There is an opinion prevalent among some of our good citizens that Broadway is not only the longest and widest, but the most superbly built, street in the world. The sooner they are undeceived the better. Broadway is a beautiful street, a very beautiful street, but it is absurd to think that our brick houses of twenty-five feet front, with plain doors and windows, built by contract in two or three months, and holding together long enough to be let, can rival the spacious stone palaces of hundreds of feet in length, with lofty gates and balconied windows, and their foundations deeply laid and slowly constructed to last for ages." This was, of course, when Broadway even below Fourteenth Street, was a residence street.

Attending service in the cathedral on Sunday, and being, as usual, wearied by the monotony and apparent insincerity of it all, he again gives vent to his feelings: —

"How admirably contrived is every part of the structure of this system to take captive the imagination. It is a religion of the imagination; all the arts of the imagination are pressed into its service; architecture, painting, sculpture, music, have lent all their charm to enchant the senses and impose on the understanding by substituting for the solemn truths of God's Word, which are addressed to the understanding, the fictions of poetry and the delusions of feeling. The theatre is a daughter of this prolific mother of abominations, and a child worthy of its dam. The lessons of morality are pretended to be taught by both, and much in the same way, by scenic effect and pantomime, and the fruits are much the same.

"I am sometimes even constrained to doubt the lawfulness of my own art when I perceive its prostitution, were I not fully persuaded that the art itself, when used for its legitimate purposes, is one of the greatest correcters of grossness and promoters of refinement. I have been led, since I have been in Italy, to think much of the propriety of introducing pictures into churches in aid of devotion. I have certainly every inducement to decide in favor of the practice did I consult alone the seeming interest of art. That pictures may and do have the effect upon some rightly to raise the affections, I have no doubt, and, abstractly considered, the practice would not merely be harmless but useful; but, knowing that man is led astray by his imagination more than by any of his other faculties, I consider it so dangerous to his best interests that I had rather sacrifice the interests of the arts, if there is any collision, than run the risk of endangering those compared with which all others are not for a moment to be considered. But more of this another time."

I have introduced here and at other times Morse's strictures on the Roman Catholic religion, and on other subjects, without comment on my part, even when these strictures seem to verge on illiberality. My desire is to present a true portrait of the man, with the shadows as well as the lights duly emphasized, fully realizing that what may appear faults to some, to others will shine out as virtues, and *vice versa.*

From Milan, Morse and his companion planned to cross the mountains to Geneva, but, having a day or two to spare, they visited the Lake of Como, which, as was to be expected, satisfied the eye of the artist: "It is shut in by mountains on either side, reminding me of the scenery of Lake George, to which its shores are very similar. In the transparency of the water, however, Lake George is its superior, and in islands also, but in all things else the Lake of Como must claim the precedence. The palaces and villas and villages which skirt its shores, the mountains, vine-clad and cultivated to their summits, all give a charm for which we look in vain as yet in our country. The luxuries of art have combined with those of nature in a wonderful degree in this enchanting spot."

On August 4, they left Milan in the diligence for Lago Maggiore, and we learn that: "Our coach is accompanied by *gendarmes.* We enquired the reason of the conductor, who was in the coach with us. He told us that the road is an unsafe one; that every day there are instances of robbery perpetrated upon those who travel alone."

It would be pleasant to follow the travellers through beautiful Maggiore and up the rugged passes from Italy to Switzerland and thence to Germany and Paris, and to see through the unspoiled eyes of an enthusiast the beauties of that playground of the nations, but it would be but the repetition of an oft-told tale, and I must hasten

HENRY CLAY

Painted by Morse. Now in the Metropolitan Museum, New York

on, making but a few extracts from the diary. No thrilling adventures were met with, except towards the end, but they enjoyed to the full the grand scenery, the picturesque costumes of the peasants and the curious customs of the different countries through which they passed. The weather was sometimes fine, but more often overcast or rainy, and we find this note on August 15: "How much do a traveller's impressions depend upon the weather, and even on the time of day in which he sees objects. He sees most of the country through which he travels but once, and it is the face which any point assumes at that one moment which is brought to his recollection. If it is under a gloomy atmosphere, it is not possible that he should remember it under other form or aspect."

On Sunday, August 28, he watched the sunrise from the summit of the Rigi under ideal conditions, and, after describing the scene and saying that the rest of the company had gone back to bed, he adds: —

"I had found too little comfort in the wretched thing that had been provided for me in the shape of a bed to desire to return thither, and I also felt too strongly the emotions which the scene I had just witnessed had excited, to wish for their dissipation in troubled dreams.

"If there is a feeling allied to devotion, it is that which such a scene of sublimity as this we have just witnessed inspires, and yet that feeling is not devotion. I am aware that it is but the emotion of taste. It may exist without a particle of true religious feeling, or it may coexist and add strength to it. There are thousands, probably, who have here had their emotion of taste excited without one thought of that Being by whom these wonders were created, one thought of their relation to Him, of their duty to Him, or of admiration at that unmerited goodness which allows them to be witnesses of his majesty and power as exhibited in these wonders of nature. Shut out as I am by circumstances from the privileges of this day in public worship, I have yet on the top of this mountain a place of private worship such as I have not had for some time past. I am alone on the mountain with such a scene spread before me that I must adore, and weak, indeed, must be that faith which, on this day, in such a scene, does not lift the heart from nature up to nature's God."

On August 30, on the road to Zurich, he makes this rather interesting observation: "We noticed in a great many instances that wires were attached to the electric rods and conducted to posts near the houses, when a chime of bells was so arranged as to ring in a highly charged state of the atmosphere (Franklin's experiment)."

Journeying on past Schaffhausen, where the beautiful falls of the Rhine filled him with admiration, he and his companion came to Heidelberg and explored the ruins of the stupendous castle. Here he parted with his travelling companion, Mr. Ferguson, who went on to Frankfort, which city Morse avoided because the French Government had established a strict quarantine against it on account of some epidemic, the nature of which is not disclosed in the notes. He was eager to get to Paris now and wished to avoid all delays.

"*September* 7. I engaged my passage in the diligence for Mannheim, and, for the first time since I have been in Europe, set out alone. . . . I learn from the gentleman in the coach that the *cordon sanitaire* in France is to be enforced with great rigor from the 11th of September; I hope, therefore, to get into France before that date.

"*September 10, Saarbruck.* We last night took our places for Metz, not knowing, however, or even thinking it probable that we should be able to get there. It was hinted by some that a small *douceur* would enable us to pass the *cordon*, but how to be applied I knew not.

"Among our passengers who joined me yesterday was a young German officer who was the only one who could speak French. With him I contrived to converse during the day. We had beds in the same room and, as we were about retiring, he told me, as I understood him, that by giving the keys of my luggage to the coachman in the morning, the business of passing at the *douane* on the frontier would be facilitated. I assented and told him, as he understood the language better than I, I left it to him to make any arrangements and I would share the expense with him.

"We were called sometime before day and I left my bed very reluctantly. The morning was cloudy and dark and so far favorable to the enterprise we were about to undertake, and of the nature and plan of which I had not the slightest suspicion. We were soon settled in the diligence and left Saarbruck for the frontier. I composed myself to sleep and had just got into a doze when suddenly the coach stopped, and, the door opening, a man touching me said in a low voice — '*Descendez, monsieur, descendez.*' I asked the reason but got no answer. My companion and I alighted. There was no house near; a bright streak in the east under the heavy black clouds showed that it was just daybreak, and ahead of us in the road a great light from the windows of a long building showed us the place of the hospital of the *cordon*.

"Our guide, for so he proved to be, taking the knapsack of my companion and a basket of mine, in which I carry my portfolio and maps, struck off to the left into a newly ploughed field, while our carriage proceeded at a quick pace onward again. I asked where we were going, but got no other reply than '*Doucement, monsieur.*' It then for the first time flashed across my mind that we had undertaken an unlawful and very hazardous enterprise, that of running by the *cordon*. I had now, however, no alternative; I must follow, for I knew not what other course to take.

"After passing through ploughed fields and wet grass and grain for some time a small by-path crossed from the main road. Our guide beckoned us back, while he went forward each way to see that all was clear, and then we crossed and proceeded again over ploughed fields and through the clover. It now began to rain which, disagreeable as it was, I did not regret, all things considered. We soon came to another and wider cross-path; we stopped and our guide went forward again in the same cautious manner, stooping down and listening, like an Indian, near the ground. He beckoned us to cross over and again we traversed the fields, passing by the base of a small hill, when, as we softly crept up the side, we saw the form of a sentinel against the light of the sky. Our guide whispered, '*Doucement*' again, and we gently retreated, my companion whispering to me, '*Très dangèreux, monsieur, très désagréable.*'

"We took a wider circuit behind some small buildings, and at length came into one of the smaller streets in the outskirts of Forbach. Here were what appeared to me barracks. The caution was given to walk softly and separately (we were all, fortunately, in dark clothes), our guide passing first round the corners, and, having passed the sentry-boxes, in which, with one exception, we saw no person, and in this instance the sentinel did not hail us (but this was in the city), we came to a house at the window of which our guide tapped. A man opened it, and, after some explanation, ascertaining who we were, opened the door and, striking a light, set some wine and bread before us.

"Here we remained for some time to recover breath after our perilous adventure, for, if one of the sentinels had seen us, we should in all probability have been instantly shot. I knew not that we were now entirely free from the danger of being arrested, until we heard our carriage in the street and had ascertained that all our luggage had passed the *douane* without suspicion. We paid our guide eight francs each, and, taking our seats again in the carriage, drove forward toward Metz."

There were no further adventures, although they trembled with anxiety every time their passports were called for. Morse regretted having been innocently led into this escapade, and would have made a clean breast of it to the police, as he had not been near Frankfort, but he feared to compromise his travelling companion who had come from that city.

On September 12 they finally arrived in Paris.

"How changed are the circumstances of this city since I was last here nearly two years ago. A traitor king has been driven into exile; blood has flowed in its streets, the price of its liberty; our friend, the nation's guest, whom I then saw at his house, with apparently little influence and out of favor with the court, the great Lafayette, is now second only to the king in honor and influence as the head of a powerful party. These and a thousand other kindred reflections, relating also to my own circumstances, crowd upon me at the moment of again entering this famous city."

CHAPTER XIX

SEPTEMBER 18, 1831 — SEPTEMBER 21, 1832

Takes rooms with Horatio Greenough. — Political talk with Lafayette. — Riots in Paris. — Letters from Greenough. — Bunker Hill Monument. — Letters from Fenimore Cooper. — Cooper's portrait by Verboeckhoven. — European criticisms. — Reminiscences of R. W. Habersham. — Hints of an electric telegraph. — Not remembered by Morse. — Early experiments in photography. — Painting of the Louvre. — Cholera in Paris. — Baron von Humboldt. — Morse presides at 4th of July dinner. — Proposes toast to Lafayette. — Letter to New York "Observer" on Fenimore Cooper. — Also on pride in American citizenship. — Works with Lafayette in behalf of Poles. — Letter from Lafayette. — Morse visits London before sailing for home. — Sits to Leslie for head of Sterne.

THE diary was not continued beyond this time and was never seriously resumed, so that we must now depend on letters to and from Morse, on fugitive notes, or on the reminiscences of others for a record of his life.

The first letter which I shall introduce was written from Paris to his brothers on September 18, 1831: —

"I arrived safely in this city on Monday noon in excellent health and spirits. My last letter to you was from Venice just as I was about to leave it, quite debilitated and unwell from application to my painting, but more, I believe, from the climate, from the perpetual sirocco which reigned uninterrupted for weeks. I have not time now to give you an account of my most interesting journey through Lombardy, Switzerland, part of Germany, and through the eastern part of France. I found, on my arrival here, my friend Mr. Greenough, the sculptor, who had come from Florence to model the bust of General Lafayette, and we are in excellent, convenient rooms together, within a few doors of the good General.

"I called yesterday on General Lafayette early in the morning. The servant told me that he was obliged to meet the Polish Committee at an early hour, and feared he could not see me. I sent in my card, however, and the servant returned immediately saying that the General wished to see me in his chamber. I followed him through several rooms and entered the chamber. The General was in dishabille, but, with his characteristic kindness, he ran forward, and, seizing both my hands, expressed with great warmth how glad he was to see me safely returned from Italy, and appearing in such good health. He then told me to be seated, and without any ceremony began familiarly to question me about my travels, etc. The conversation, however, soon turned upon the absorbing topic of the day, the fate of Poland, the news of the fall of Warsaw having just been received by telegraphic dispatch. I asked him if there was now any hope for Poland. He replied: 'Oh, yes! Their cause is not yet desperate; their army is safe; but the conduct of France, and more especially of England, has been most pusillanimous and culpable. Had the English Government shown the least disposition to coalesce in vigorous measures with France for the assistance of the Poles, they would have achieved their independence.'

"The General looks better and younger than ever. There is a healthy freshness of complexion, like that of a young man in full vigor, and his frame and step (allowing for his lameness) are as firm and strong as when he was our nation's guest. I sat with him ten or fifteen minutes and then took my leave, for I felt it a sin to consume any more of the time of a man engaged as he is in great plans of benevolence, and whose every moment is, therefore, invaluable.

"The news of the fall of Warsaw is now agitating Paris to a degree not known since the trial of the ex-ministers. About three o'clock our servant told us that there was fighting at the Palais Royal, and we determined to go as far as we prudently could to see the tumult. We proceeded down the Rue Saint-Honoré. There was evident agitation in the multitudes that filled the sidewalks — an apprehension of something to be dreaded. There were groups at the corners; the windows were filled, persons looking out as if in expectation of a procession or of some fête. The shops began to be shut, and every now and then the drum was heard beating to arms. The troops were assembling and bodies of infantry and cavalry were moving through the various streets. During this time no noise was heard from the people — a mysterious silence was observed, but they were moved by the slightest breath. If one walked quicker than the rest, or suddenly stopped, thither the enquiring look and step were directed, and a group instantly assembled. At the Palais Royal a larger crowd had collected and a greater body of troops were marching and countermarching in the Place du Palais Royal. The Palais Royal itself had the interior cleared and all the courts. Everything in this place of perpetual gayety was now desolate; even the fountains had ceased to play, and the seared autumnal leaves of the trees, some already fallen, seemed congruous with the sentiment of the hour. Most of the shops were also shut and the stalls deserted. Still there was no outcry and no disturbance.

"Passing through the Rue Vivienne the same collections of crowds and of troops were seen. Some were reading a police notice just posted on the walls, designed to prevent the riotous assembling of the people, and advising them to retire when the riot act should be read. The notice was read with murmurs and groans, and I had scarcely ascertained its contents before it was torn from the walls with acclamations. As night approached we struck into the Boulevard de la Madeleine. At the corner of this boulevard and the Rue des Capucines is the hotel of General Sebastiani. We found before the gates a great and increasing crowd.

"We took a position on the opposite corner, in such a place as secured a safe retreat in case of need, but allowed us to observe all that passed. Here there was an evident intention in the crowd of doing some violence, nor was it at all doubtful what would be the object of their attack. They seemed to wait only for the darkness and for a leader.

"The sight of such a crowd is fearful, and its movements, as it was swayed by the incidents of the moment, were in the highest degree exciting. A body of troops of the line would pass; the crowd would silently open for their passage and close immediately behind them. A body of the National Guard would succeed, and these would be received with loud cheers and gratulations. A soldier on guard would exercise a little more severity than was, perhaps, necessary for the occasion; yells, and execrations, and hisses would be his reward.

"Night had now set in; heavy, dark clouds, with a misty rain, had made the heavens above more dark and gloomy. A man rushed forward toward the gate, hurling his hat in the air, and followed by the crowd, which suddenly formed into long lines behind him. I now looked for something serious. A body of troops was in line before the gate. At this moment two police officers, on horseback, in citizens' dress, but with a tricolored belt around their bodies, rode through the crowd and up to the gate, and in a moment after I perceived the multitude from one of the streets rushing in wild confusion into the boulevard, and the current of the people setting back in all directions.

"While wondering at the cause of this sudden movement, I heard the trampling of horses, and a large band of carabiniers, with their bright helmets glittering in the light of the lamps, dashed down the street and drew up before the gate. The police officers put themselves at their head and harangued the people. The address was received with groans. The *carabiniers* drew their swords, orders were given for the charge, and in an instant they dashed down the street, the people dispersing like the mist before the wind. The charge was made down the opposite sidewalk from that where we had placed ourselves, so I kept my station, and, when they returned up the middle of the street to charge on the other side, I crossed over behind them and avoided them."

I have given enough of this letter to show that Morse was still surrounded by dangers of various sorts, and it is also a good pen-picture of the irresponsible actions of a cowardly mob, especially of a Parisian mob.

The letters which passed between Morse and his friends, James Fenimore Cooper, the novelist, and Horatio Greenough, the sculptor, are most interesting, and would of themselves fill a volume. Both Cooper and

Greenough wrote fluently and entertainingly, and I shall select a few characteristic sentences from the letters of each, resisting the strong temptation to include the whole correspondence.

Greenough returned to Florence after having roomed with Morse in Paris, and wrote as follows from there: —

As for the commission from Government, I don't speak of it yet. After about a fortnight I shall be calm, I think. Morse, I have made up my mind on one score, namely, that this order shall not be fruitless to the greater men who are now in our rear. They are sucking now and rocking in cradles, but I can hear the pung! pung! puffetty! of their hammers, and I am prophetic, too. We'll see if Yankee land can't muster some ten or a dozen of them in the course of as many years. . . .

You were right, I had heard of the resolution submitted to Congress, etc. Mr. Cooper wrote me about it. I have not much faith in Congress, however. I will confess that, when the spectre Debt has leaned over my pillow of late, and, smiling ghastlily, has asked if she and I were not intended as companions through life, I snap my fingers at her and tell her that Brother Jonathan talks of adopting me, and that he won't have her of his household. "Go to London, you hag," says I, "where they say you're handsome and wholesome; don't grind your long teeth at me, or I'll read the Declaration of Independence to ye." So you see I make uncertain hopes fight certain fears, and borrow from the generous, good-natured Future the motives for content which are denied me by the stinted Present. . . .

What shall I say in answer to your remarks on my

opinions? Shall I go all over the ground again? It were useless. That my heart is wrong in a thousand ways I daily feel, but 't is my stubborn head which refuses to comprehend the creation as you comprehend it. That we should be grateful for all we have, I feel — for all we have is given us; nor do I think we have little. For my part I would be blest in mere existence were I not goaded by a wish to make my one talent two; and we have Scripture for the rectitude of such a wish. I don't think the stubborn resistance of the tide of ill-fortune can be called rebellion against Providence. "Help-yourself and Heaven will help you," says the proverb. . . .

There hangs before me a print of the Bunker Hill Monument. Pray be judge between me and the building committee of that monument. There you observe that my model was founded solidly, and on each of its square plinths were trophies, or groups, or cannon, as might be thought fit. (No. I.)

Well, they have taken away the foundation, made the shaft start sheer from the dirt like a spear of asparagus, and, instead of an acute angle, by which I hoped to show the work was done and lead off the eye, they have made an obtuse one, producing the broken-chimney-like effect which your eye will not fail to condemn in No. II. Then they have enclosed theirs with a light, elegant fence, *à la Parigina*, as though the austere forms of Egypt were compatible with the decorative flummery of the boulevards. Let 'em go for dunderheads as they are. . . .

I congratulate you on your sound conscience with regard to the affair that you wot of. As for your remaining free, that's all very well to think during the interregnum, but a man without a true love is a ship without ballast, a one-tined fork, half a pair of scissors, an utter flash in the pan. . . . So you are going home, my dear Morse, and God knows if ever I shall see you again. Pardon, I pray you, anything of levity which you may have been offended at in me. Believe me it arose from my so rarely finding one to whom I could be natural and give loose without fear of good faith or good nature ever failing. Wherever I am your approbation will be dearer to me than the hurrah of a world. I shall write to glorious Fenimore in a few days. My love to Allston and Dana. God bless you,

H. GREENOUGH.

These extracts are from different letters, but they show, I think, the charming character of the man and reflect his admiration for Morse. From the letters of James Fenimore Cooper, written while they were both in Europe, I select the two following as characteristic:

July 31, 1832.

MY DEAR MORSE, — Here we are at Spa — the famous hard-drinking, dissipated, gambling, intriguing Spa — where so much folly has been committed, so many fortunes squandered, and so many women ruined! How are the mighty fallen! We have just returned from a ramble in the environs, among deserted reception-houses and along silent roads. The country is not unlike Ballston, though less wooded, more cultivated, and perhaps a little more varied. . . . I have had a great compliment paid me, Master Samuel, and, as it is nearly the only compliment I have received in travelling over Europe, I am the more proud of it. Here are the facts.

You must know there is a great painter in Brussels of

the name of Verboeckhoven (which, translated into the vernacular, means a *bull and a book baked in an oven!*), who is another Paul Potter. He outdoes all other men in drawing cattle, etc., with a suitable landscape. In his way he is truly admirable. Well, sir, this artist did me the favor to call at Brussels with the request that I would let him sketch my face. He came after the horses were ordered, and, knowing the difficulty of the task, I thanked him, but was compelled to refuse. On our arrival at Liège we were told that a messenger from the Governor had been to enquire for us, and I began to bethink me of my sins. There was no great cause for fear, however, for it proved that Mr. Bull-and-book-baked had placed himself in the diligence, come down to Liège (sixty-three miles), and got the Governor to give him notice, by means of my passport, when we came. Of course I sat.

I cannot say the likeness is good, but it has a vastly life-like look and is like all the other pictures you have seen of my chameleon face. Let that be as it will, the compliment is none the less, and, provided the artist does not mean to serve me up as a specimen of American wild beasts, I shall thank him for it. To be followed twelve posts by a first-rate artist, who is in favor with the King, is so unusual that I was curious to know how far our minds were in unison, and so I probed him a little. I found him well skilled in his art, of course, but ignorant on most subjects. As respects our general views of men and things there was scarcely a point in common, for he has few salient qualities, though he is liberal; but his gusto for natural subjects is strong, and his favorite among all my books is "The Prairie," which, you know, is filled with wild beasts. Here the secret was out. That picture of animal nature had so caught his fancy that he followed me sixty miles to paint a sketch.

While this letter of Cooper's was written in lighter vein, the following extracts from one written on August 19 show another side of his character: —

The criticisms of which you speak give me no concern. . . . The "Heidenmauer" is not equal to the "Bravo," but it is a good book and better than two thirds of Scott's. They may say it is like his if they please; they have said so of every book I have written, even the "Pilot." But the "Heidenmauer" is like and was intended to be like, in order to show how differently a democrat and an aristocrat saw the same thing. As for French criticisms they have never been able to exalt me in my own opinion nor to stir my bile, for they are written with such evident ignorance (I mean of English books) as to be beneath notice. What the deuce do I care whether my books are on their shelves or not? What did I ever get from France or Continental Europe? Neither personal favors nor money. But this they cannot understand, for so conceited is a Frenchman that many of them think that I came to Paris to be paid. Now I never got the difference in the boiling of the pot between New York and Paris in my life. The "Journal des Débats" was snappish with "Water Witch," merve [?] I believe with "Bravo," and let it bark at "Heidenmauer" and be hanged.

No, no more. The humiliation comes from home. It is biting to find that accident has given me a country which has not manliness and pride to maintain its own opinions, while it is overflowing with conceit. But never mind all this. See that you do not decamp before my departure and I'll promise not to throw myself into the Rhine. . . .

I hope the Fourth of July is not breaking out in Habersham's noddle, for I can tell him that was the place most affected during the dinner. Adieu,

Yours as ever,
J. Fenimore Cooper.

The Mr. Habersham here jokingly referred to was R. W. Habersham, of Augusta, Georgia, who in the year 1831 was an art student in the *atelier* of Baron Gros, and between whom and Morse a friendship sprang up. They roomed together at a time when the cholera was raging in Paris, but, owing to Mr. Habersham's wise insistence that all the occupants of the house should take a teaspoonful of charcoal every morning, all escaped the disease.

Mr. Habersham in after years wrote and sent to Morse some of his reminiscences of that period, and from these I shall quote the following as being of more than ordinary interest: —

"The Louvre was always closed on Monday to clean up the gallery after the popular exhibition of the paintings on Sunday, so that Monday was our day for visits, excursions, etc. On one occasion I was left alone, and two or three times during the week he was absent. This was unusual, but I asked no questions and made no remarks. But on Saturday evening, sitting by our evening lamp, he seemed lost in thought, till suddenly he remarked: 'The mails in our country are too slow; this French telegraph is better, and would do even better in our clear atmosphere than here, where half the time fogs obscure the skies. But this will not be fast enough — *the lightning would serve us better.*'

"These may not be the exact words, but they convey the sense, and I, laughing, said: 'Aha! I see what you have been after, you have been examining the French system of telegraphing.' He admitted that he had taken advantage of the kind offer of one in authority to do so. . . .

"There was, on one occasion, another reference made to the conveyance of sound under water, and to the length of time taken to communicate the letting in of the water into the Erie Canal by cannon shots to New York, and other means, during which the suggestion of using keys and wires, like the piano, was rejected as requiring too many wires, if other things were available. I recollect also that in our frequent visits to Mr. J. Fenimore Cooper's, in the Rue St. Dominique, these subjects, so interesting to Americans, were often introduced, and that Morse seemed to harp on them, constantly referring to Franklin and Lord Bacon. Now I, while recognizing the intellectual grandeur of both these men, had contracted a small opinion of their moral strength; but Morse would uphold and excuse, or rather deny, the faults attributed. Lord Bacon, especially, he held to have *sacrificed himself to serve the queen in her aberrations;* while of Franklin, 'the Great American,' recognized by the French, he was particularly proud."

Cooper also remembered some such hints of a telegraph made by Morse at that time, for in "The Sea Lions,"[1] on page 161, he says: —

[1] The Riverside Press, 1870.

"We pretend to no knowledge on the subject of the dates of discoveries in the arts and sciences, but well do we remember the earnestness, and single-minded devotion to a laudable purpose, with which our worthy friend first communicated to us his ideas on the subject of using the electric spark by way of a telegraph. It was in Paris and during the winter of 1831–32 and the succeeding spring, and we have a satisfaction in recording this date that others may prove better claims if they can."

Curiously enough, Morse himself could, in after years, never remember having suggested at that time the possibility of using electricity to convey intelligence. He always insisted that the idea first came to him a few months later on his return voyage to America, and in 1849 he wrote to Mr. Cooper saying that he must be mistaken, to which the latter replied, under date of May 18: —

"For the time I still stick to Paris, so does my wife, so does my eldest daughter. You did no more than to throw out the general idea, but I feel quite confident this occurred in Paris. I confess I thought the notion evidently chimerical, and as such spoke of it in my family. I always set you down as a sober-minded, common-sense sort of a fellow, and thought it a high flight for a painter to make to go off on the wings of the lightning. We may be mistaken, but you will remember that the priority of the invention was a question early started, and my impressions were the same much nearer to the time than it is to-day."

That the recollections of his friends were probably clearer than his own on this point is admitted by Morse in the following letter: —

IRVING HOUSE,
NEW YORK, September 5, 1849.

MY DEAR SIR, — I was agreeably surprised this morning in conversing with Professor Renwick to find that he corroborates the fact you have mentioned in your "Sea Lions" respecting the earlier conception of my telegraph by me, than the date I had given, and which goes only so far back in my own recollection as 1832. Professor Renwick insists that immediately after Professor Dana's lectures at the New York Athenæum, I consulted with him on the subject of the velocity of electricity and in such a way as to indicate to him that I was contriving an electric telegraph. The consultation I remember, but I did not recollect the time. He will depose that it was before I went to Europe, after those lectures; now I went in 1829; this makes it almost certain that the impression you and Mrs. Cooper and your daughter had that I conversed with you on the subject in 1831 after my return from Italy is correct.

If you are still persuaded that this is so, your deposition before the Commission in this city to that fact will render me an incalculable service. I will cheerfully defray your expenses to and from the city if you will meet me here this week or beginning of next.

In haste, but with best respects to Mrs. Cooper and family,

I am, dear sir, as ever your friend and servant,
SAML. F. B. MORSE.

J. FENIMORE COOPER, ESQ.

All this is interesting, but, of course, has no direct bearing on the actual date of invention. It is more than probable that Morse did, while he was studying the French semaphores, and at an even earlier date, dream vaguely of the possibility of using electricity for conveying intelligence, and that he gave utterance among his intimates to these dreams; but the practical means of so utilizing this mysterious agent did not take shape in his mind until 1832. An inchoate vision of the possibility of using electricity is far different from an actual plan eventually elaborated into a commercial success.

Another extract from Mr. Habersham's reminiscences, on a totally different subject, will be found interesting: "I have forgot to mention that one day, while in the Rue Surenne, I was studying from my own face reflected in a glass, as is often done by young artists, when I remarked how grand it would be if we could invent a method of fixing the image on the mirror. Professor Morse replied that he had thought of it while a pupil at Yale, and that Professor Silliman (I think) and himself had tried it with a wash of nitrate of silver on a piece of paper, but that, unfortunately, it made the lights *dark* and the shadows *light*, but that if they could be reversed, we should have a facsimile like India-ink drawings. Had they thought of using glass, as is now done, the daguerreotype would have been perhaps anticipated — certainly the photograph."

This is particularly interesting because, as I shall note later on, Morse was one of the pioneers in experimenting with the daguerreotype in America.

Among the paintings which Morse executed while he was in Paris was a very ambitious one. This was an interior of one of the galleries in the Louvre with carefully executed miniature copies of some of the most celebrated canvases. Writing of it, and of the dreadful epidemic of cholera, to his brothers on May 6, 1832, he says: —

"My anxiety to finish my picture and to return drives me, I fear, to too great application and too little exercise, and my health has in consequence been so deranged that I have been prevented from the speedy completion of my picture. From nine o'clock until four daily I paint uninterruptedly at the Louvre, and, with the closest application, I shall not be able to finish it before the close of the gallery on the 10th of August. The time each morning before going to the gallery is wholly employed in preparation for the day, and, after the gallery closes at four, dinner and exercise are necessary, so that I have no time for anything else.

"The cholera is raging here, and I can compare the state of mind in each man of us only to that of soldiers in the heat of battle; all the usual securities of life seem to be gone. Apprehension and anxiety make the stoutest hearts quail. Any one feels, when he lays himself down at night, that he will in all probability be attacked before daybreak; for the disease is a pestilence that walketh in darkness, and seizes the greatest number of its victims at the most helpless hour of the night. Fifteen hundred were seized in a day, and fifteen thousand at least have already perished, although the official accounts will not give so many.

"*May 14.* My picture makes progress and I am sanguine of success if nothing interferes to prevent its completion. I shall take no more commissions here and shall only complete my large picture and a few unfinished works.

"General Lafayette told me a few weeks ago, when I was returning with him in his carriage, that the financial condition of the United States was a subject of great importance, and he wished that I would write you and others, who were known as statistical men, and get your views on the subject. There never was a better time for demonstrating the principles of our free institutions by showing a result favorable to our country."

Among the men of note whom Morse met while he was in Paris was Baron Alexander von Humboldt, the famous traveller and naturalist, who was much attracted towards the artist, and often went to the Louvre to watch him while he was at work, or to wander through the galleries with him, deep in conversation. He was afterwards one of the first to congratulate Morse on the successful exhibition of his telegraph before the French Academy of Science.

As we have already seen, Morse was intensely patriotic. He followed with keen interest the developments in our national progress as they unrolled themselves before his eyes, and when the occasion offered he took active part in furthering what he considered the right and in vigorously denouncing the wrong. He was never blind to our national or party failings, but held the mirror up before his countrymen's eyes with steady hand, and yet he was prouder of being an American than of anything else, and, as I have had occasion to remark before, his ruling passion was an intense desire to accomplish some great good for his beloved country, to raise her in the estimation of the rest of the world.

On the 4th of July, 1832, he was called on to preside at the banquet given by the Americans resident in Paris, with Mr. Cooper as vice-president. General Lafayette was the guest of honor, and the American Minister Hon. William C. Rives, G. W. Haven, and many others were present.

Morse, in proposing the toast to General Lafayette, spoke as follows: —

"I cannot propose the next toast, gentlemen, so intimately connected with the last, without adverting to the distinguished honor and pleasure we this day enjoy above the thousands, and I may say hundreds of thousands, of our countrymen who are at this moment celebrating this great national festival — the honor and pleasure of having at our board our venerable guest on my right hand, the hero whom two worlds claim as their own. Yes, gentlemen, he belongs to America as well as to Europe. He is our fellow citizen, and the universal voice of our country would cry out against us did we not manifest our nation's interest in his person and character.

"With the mazes of European politics we have nothing to do; to changing schemes of good or bad government we cannot make ourselves a party; with the success or defeat of this or that faction we can have no sympathy; but with the great principles of rational liberty, of civil and religious liberty, those principles for which our guest fought by the side of our fathers, and which he has steadily maintained for a long life, 'through good report and evil report,' we do sympathize. We should not be Americans if we did not sympathize with them, nor can we compromise one of these principles and preserve our self-respect as loyal American citizens. They are the principles of order and good government, of obedience to law; the principles which, under Providence, have made our country unparalleled in prosperity; principles which rest, not in visionary theory, but are made palpable by the sure test of experiment and time.

"But, gentlemen, we honor our guest as the stanch, undeviating defender of these principles, of our principles, of American principles. Has he ever deserted them? Has he ever been known to waver? Gentlemen, there are some men, some, too, who would wish to direct public opinion, who are like the buoys upon tide-water. They float up and down as the current sets this way or that. If you ask at an emergency where they are, we cannot tell you; we must first consult the almanac; we must know the quarter of the moon, the way of the wind, the time of the tide, and then we may guess where you will find them.

"But, gentlemen, our guest is not of this fickle class. He is a tower amid the waters, his foundation is upon a rock, he moves not with the ebb and flow of the stream. The storm may gather, the waters may rise and even dash above his head, or they may subside at his feet, still he stands unmoved. We know his site and his bearings, and with the fullest confidence we point to where he stood six-and-fifty years ago. He stands there now. The winds have swept by him, the waves have dashed around him, the snows of winter have lighted upon him, but still he is there.

"I ask you, therefore, gentlemen, to drink with me in honor of General Lafayette."

Portions of many of Morse's letters to his brothers were published in the New York "Observer," owned and edited by them. Part of the following letter was so published, I believe, but, at Mr. Cooper's request, the sentences referring to his personal sentiments were omitted. There can be no harm, however, in giving them publicity at this late day.

The letter was written on July 18, 1832, and begins by gently chiding his brothers for not having written to him for nearly four months, and he concludes this part by saying, "But what is past can't be helped. I am glad, exceedingly glad, to hear of your prosperity and hope it may be continued to you." And then he says: —

"I am diligently occupied every moment of my time at the Louvre finishing the great labor which I have there undertaken. I say 'finishing,' I mean that part of it which can only be completed there, namely, the copies of the pictures. All the rest I hope to do at home in New York, such as the frames of the pictures, the figures, etc. It is a great labor, but it will be a splendid and valuable work. It excites a great deal of attention from strangers and the French artists. I have many compliments upon it, and I am sure it is the most *correct* one of *its kind* ever painted, for every one says I have caught the style of each of the masters. Cooper is delighted with it and I think he will own it. He is with me two or three hours at the gallery (the hours of his relaxation) every day as regularly as the day comes. I spend almost every evening at his house in his fine family.

"Cooper is very little understood, I believe, by our good people. He has a bold, original, independent mind, thoroughly American. He loves his country and her principles most ardently; he knows the hollowness of all the despotic systems of Europe, and especially is he thoroughly conversant with the heartless, false, selfish system of Great Britain; the perfect antipodes of our own. He fearlessly supports American principles in the face of all Europe, and braves the obloquy and intrigues against him of all the European powers. I say all the European powers, for Cooper is more read, and, therefore, more feared, than any American, — yes, more than any European with the exception, possibly, of Scott. His works are translated into all the languages of the Continent; editions of every work he publishes are printed in, I think, more than thirty different cities, and all this without any pains on his part. He deals, I believe, with only one publisher in Paris and one in London. He never asks what effect any of his sentiments will have upon the sale of his works; the only question he asks is — 'Are they just and true?'

"I know of no man, short of a true Christian, who is so truly guided by high principles as Cooper. He is not a religious man (I wish from my heart he was), yet he is theoretically orthodox, a great respecter of religion and religious men, a man of unblemished moral character. He is courted by the greatest and the most aristocratic, yet he never compromises the dignity of an American citizen, which he contends is the highest distinction a man can have in Europe, and there is not a doubt but he commands the respect of the exclusives here in a tenfold degree more than those who truckle and cringe to European opinions and customs. They love an independent man and know enough of their own heartless system to respect a real freeman. I admire exceedingly his proud assertion of the rank of an American (I speak from a political point of view), for I know no reason why an American should not take rank, and assert it, too, above any of the artificial distinctions that Europe has made. We have no aristocratic grades, no titles of nobility, no ribbons, and garters, and crosses, and other gewgaws that please the great babies of Europe; are we, therefore, to take rank below or above them? I say above them, and I hope that every American who comes abroad will feel that he is bound, for his country's sake, to take that stand. I don't mean ostentatiously, or offensively, or obtrusively, but he ought to have an American self-respect.

"There can be no *condescension* to an American. An American gentleman is equal to any title or rank in Europe, kings and emperors not excepted. Why is he not? By what law are we bound to consider ourselves inferior because we have stamped *folly* upon the artificial and unjust grades of European systems, upon these antiquated remnants of feudal barbarism?

"Cooper sees and feels the absurdity of these distinctions, and he asserts his American rank and maintains it, too, I believe, from a pure patriotism. Such a man deserves the support and respect of his countrymen, and I have no doubt he has them. . . . It is high time we should assume a more American tone while Europe is leaving no stone unturned to vilify and traduce us, because the rotten despotisms of Europe fear our example and hate us. You are not aware, perhaps, that the *Trollope* system is political altogether. You think that, because we know the grossness of her libels and despise her abuse, England and Europe do the same. You are mistaken; they wish to know no good of us. Mrs. Trollope's book is more popular in England (and that, too, among a class who you fain would think know better) than any book of travels ever published in America.[1] It is also translating into French, and will be puffed and extolled by France, who is just entering upon the system of vilification of America and her institutions, that England has been pursuing ever since we as colonies resisted her oppressive measures. Tory England, aristocratic England, is the same now towards us as she was then, and Tory France, aristocratic France, follows in her steps. We may deceive ourselves on this point by knowing the kindly feeling manifested by religious and benevolent men towards each other in both countries, but we shall be wanting in our usual Yankee penetration if the good feeling of these excellent and pious men shall lead us to think that their governments, or even the mass of their population, are actuated by the same kindly regard. No, they hate us, cordially hate us. We should not disguise the truth, and I will venture to say that no genuine American, one who loves his country and her distinctive principles, can live abroad in any of the countries of Europe, and not be thoroughly convinced that Europe, as it is, and America, as it is, can have no feeling of cordiality for each other.

"America is the stronghold of the popular principle, Europe of the despotic. These cannot unite; there can be, at present, no sympathy. . . . We need not quarrel with Europe, but we must keep ourselves aloof and suspect all her manœuvres. She has no good will towards us and we must not be duped by her soft speeches and fair words, on the one side, nor by her contemptible detraction on the other."

[1] This refers to Mrs. Frances Trollope's book *Domestic Manners of the Americans*, which created quite a stir in its day.

Morse found time, in spite of his absorption in his artistic work, to interest himself and others in behalf of the Poles who had unsuccessfully struggled to maintain their independence as a nation. He was an active member of a committee organized to extend help to them, and this committee was instrumental in obtaining the release from imprisonment in Berlin of Dr. S. G. Howe, who "had been entrusted with twenty thousand francs for the relief of the distressed Poles." In this work he was closely associated with General Lafayette, already his friend, and their high regard for each other was further strengthened and resulted in an interchange of many letters. Some of these were given away by Morse to friends desirous of possessing autographs of the illustrious Lafayette; others are still among his papers, and some of these I shall introduce in their proper chronological order. The following one was written on September 27, 1832, from La Grange: —

My dear Sir, — I am sorry to see you will not take Paris and La Grange in your way to Havre, unless you were to wait for the packet of the 10th in company with General Cadwalader, Commodore Biddle, and those young, amiable Philadelphians who contemplate sailing on that day. But if you persist to go by the next packet, I beg you here to receive my best wishes and those of my family for your happy voyage.

Upon you, my dear sir, I much depend to give our friends in the United States a proper explanation of the state of things in Europe. You have been very attentive to what has passed since the Revolution of 1830. Much has been obtained here and in other parts of Europe in this whirlwind of a week. Further consequences here and in other countries — Great Britain and Ireland included — will be the certain result, though they have been mauled and betrayed where they ought to have received encouragement. But it will not be so short and so cheap as we had a right to anticipate it might be. I think it useful, on both sides of the water, to dispel the cloud which ignorance or design may throw over the real state of European and French politics.

In the mean while I believe it to be the duty of every American returned home to let his fellow citizens know what wretched handle is made of the violent collisions, threats of a separation, and reciprocal abuse, to injure the character and question the stability of republican institutions. I too much depend upon the patriotism and good sense of the several parties in the United States to be afraid that those dissensions may terminate in a final dissolution of the Union; and should such an event be destined in future to take place, deprecated as it has been by the best wishes of the departed founders of the Revolution, — Washington at their head, — it ought at least, in charity, not to take place before the not remote period when every one of those who have fought and bled in the cause shall have joined their contemporaries.

What is to be said of Poland and the situation of her heroic, unhappy sons, you well know, having been a constant and zealous member of our committee.

You know what sort of mental perturbation, among the ignorant part of every European nation, has accompanied the visit of the cholera in Russia, Germany, Hungary, and several parts of Great Britain and France — suspicions of poison, prejudices against the politicians, and so forth. I would like to know whether the population of the United States has been quite free of these aberrations, as it would be an additional argument in behalf of republican institutions and superior civilization resulting from them.

Most truly and affectionately,
Your friend,
LAFAYETTE.

As we see from the beginning of this letter, Morse had now determined to return home. He had executed all the commissions for copies which had been given to him, and his ambitious painting of the interior of the Louvre was so far finished that he could complete it at home. He sailed from Havre on the 1st of October in the packet-ship Sully. The name of this ship has now become historic, and a chance conversation in mid-ocean was destined to mark an epoch in human evolution. Before sailing, however, he made a flying trip to England, and he writes to his brothers from London on September 21: —

"Here I am once more in England and on the wing *home*. I shall probably sail from Havre in the packet of October 1 (the Sully), and I shall leave London for Southampton and Havre on the 26th inst., to be prepared for sailing.

"I am visiting old friends and renewing old associations in London. Twenty years make a vast difference as well in the aspect of this great city as in the faces of old acquaintances. London may be said literally to have gone into the country. Where I once was accustomed to walk in the fields, so far out of town as even to shoot at a target against the trees with impunity, now there are spacious streets and splendid houses and gardens.

"I spend a good deal of my spare time with Leslie. He is the same amiable, intelligent, unassuming gentleman that I left in 1815. He is painting a little picture — 'Sterne recovering his Manuscripts from the Curls of his Hostess at Lyons.' I have been sitting to him for the head of Sterne, whom he thinks I resemble very strongly. At any rate, he has made no alteration in the character of the face from the one he had drawn from Sterne's portrait, and has simply attended to the expression.

"When I left Paris I was feeble in health, so much so that I was fearful of the effects of the journey to London, especially as I passed through villages suffering severely from the cholera. But I proceeded moderately, lodged the first night at Boulogne-sur-Mer, crossed to Dover in a severe southwest gale, and passed the next night at Canterbury, and the next day came to London. I think the ride did me good, and I have been exercising a great deal, riding and walking, since, and my general health is certainly improving. I am in hopes that the voyage will completely set me up again."

CHAPTER XX

Morse's life almost equally divided into two periods, artistic and scientific. — Estimate of his artistic ability by Daniel Huntington. — Also by Samuel Isham. — His character as revealed by his letters, notes, etc. — End of Volume I.

MORSE's long life (he was eighty-one when he died) was almost exactly divided, by the nature of his occupations, into two equal periods. During the first, up to his forty-first year, he was wholly the artist, enthusiastic, filled with a laudable ambition to excel, not only for personal reasons, but, as appears from his correspondence, largely from patriotic motives, from a wish to rescue his country from the stigma of pure commercialism which it had incurred in the eyes of the rest of the world. It is true that his active brain and warm heart spurred him on to interest himself in many other things, in inventions of more or less utility, in religion, politics, and humanitarian projects; but next to his sincere religious faith, his art held chiefest sway, and everything else was made subservient to that.

During the latter half of his life, however, a new goddess was enshrined in his heart, a goddess whose cult entailed even greater self-sacrifice; keener suffering, both mental and physical; more humiliation to a proud and sensitive soul, shrinking alike from the jeers of the incredulous and the libels and plots of the envious and the unscrupulous.

While he plied his brush for many years after the conception of his epoch-making invention, it was with an ever lessening enthusiasm, with a divided interest. Art no longer reigned supreme; Invention shared the throne with her and eventually dispossessed her. It seems, therefore, fitting that, in closing the chronicle of Morse the artist, his rank in the annals of American art should be estimated as viewed by a contemporary and by the more impartial historian of the present day.

From a long article prepared by the late Daniel Huntington for Mr. Prime, I shall select the following passages: —

"My acquaintance with Professor Morse began in the spring of 1835, when I was placed under his care by my father as a pupil. He then lived in Greenwich Lane (now Greenwich Avenue), and several young men were studying art under his instruction. . . . He gave a short time every day to each pupil, carefully pointing out our errors and explaining the principles of art. After drawing for some time from casts with the crayon, he allowed us to begin the use of the brush, and we practised painting our studies from the casts, using black, white, and raw umber.

"I believe this method was of great use in enabling us early to acquire a good habit of painting. I only regret that he did not insist on our sticking to this kind of study a longer time and drill us more severely in it; but he indulged our hankering for color too soon, and, when once we had tasted the luxury of a full palette of colors, it was a dry business to go back to plain black and white.

"In the autumn of that year, 1835, he removed to spacious rooms in the New York University on Washington Square. In the large studio in the north wing he painted several fine portraits, among them the beautiful full-length of his daughter, Mrs. Lind. He also lectured before the students and a general audience, illustrating his subject by painted diagrams. . . .

"Professor Morse's love of scientific experiments was shown in his artist life. He formed theories of color, tried experiments with various vehicles, oils, varnishes, and pigments. His studio was a kind of laboratory. A beautiful picture of his wife and two children was painted, he told me, with colors ground in milk, and the effect was juicy, creamy, and pearly to a degree. Another picture was commenced with colors mixed with beer; afterwards solidly impasted and glazed with rich, transparent tints in varnish. His theory of color is fully explained in the account of his life in Dunlap's 'Arts of Design.' He proved its truth by boxes and balls of various colors. He had an honest, solid, vigorous *impasto*, which he strongly insisted on in his instructions — a method which was like the great masters of the Venetian school. This method was modified in his practice by his studies under West in England, and by his intimacy with Allston, for whose genius he had a great reverence, and by whose way of painting he was strongly influenced.

"He was a lover of simple, unaffected truth, and this trait is shown in his works as an artist. He had a passion for color, and rich, harmonious tints run through his pictures, which are glowing and mellow, and yet pearly and delicate.

"He had a true painter's eye, but he was hindered from reaching the fame his genius promised as a painter by various distractions, such as the early battles of the Academy of Design in its struggles for life, domestic afflictions, and, more than all, the engrossing cares of his invention.

SUSAN W. MORSE

Eldest daughter of the artist

"The 'Hercules,' with its colossal proportions and daring attitude, is evidence of the zeal and courage of his early studies. . . . It is worthy of being carefully preserved in a public gallery, not only as an instance of successful study in a young artist (Morse was in his twenty-first year), but as possessing high artistic merit, and a force and richness which plainly show that, if his energies had not been diverted, he might have achieved a name in art equal to the greatest of his contemporaries. . . .

"Professor Morse's world-wide fame rests, of course, on his invention of the electric telegraph; but it should be remembered that the qualities of mind which led to it were developed in the progress of his art studies, and if his paintings, in the various fields of history, portrait, and landscape, could be brought together, it would be found that he deserved an honored place among the foremost American artists."

This was an estimate of Morse's ability as a painter by a man of his own day, a friend and pupil. As this would, naturally, be somewhat biased, it will be more to the point to see what a competent critic of the present day has to say.

Mr. Samuel Isham, in his authoritative "History of American Painting," published in 1910, after giving a brief biographical sketch of Morse and telling why he came to abandon the brush, thus sums up: —

"It was a serious loss, for Morse, without being a genius, was yet, perhaps, better calculated than another to give in pictures the spirit of the difficult times from 1830 to 1860. He was a man sound in mind and body, well born, well educated, and both by birth and education in sympathy with his time. He had been abroad, had seen good work, and received sound training. His ideals were not too far ahead of his public. Working as he did under widely varying conditions, his paintings are dissimilar, not only in merit but in method of execution; even his portraits vary from thin, free handling to solid *impasto*. Yet in the best of them there is a real painter's feeling for his material; the heads have a soundness of construction and a freshness in the carnations that recall Raeburn rather than West; the poses are graceful or interesting, the costumes are skilfully arranged, and in addition he understands perfectly the character of his sitters, the men and women of the transition period, shrewd, capable, but rather commonplace, without the ponderous dignity of Copley's subjects or the cosmopolitan graces of a later day.

"The struggles incident to the invention and development of telegraphy turned Morse from the practice of art, but up to the end of his life he was interested in it and aggressive in any scheme for its advancement."

I think that from the letters, notes, etc., which I have in the preceding pages brought together, a clear conception of Morse's character can be formed. The dominant note was an almost childlike religious faith; a triumphant trust in the goodness of God even when his hand was wielding the rod; a sincere belief in the literal truth of the Bible, which may seem strange to us of the twentieth century; a conviction that he was destined in some way to accomplish a great good for his fellow men.

Next to love of God came love of country. He was patriotic in the best sense of the word. While abroad he stoutly upheld the honor of his native land, and at home he threw himself with vigor into the political discussions of the day, fighting stoutly for what he considered the right. While sometimes, in the light of future events, he seems to have erred in allowing his religious beliefs to tinge too much his political views, he was always perfectly sincere and never permitted expediency to brush aside conviction.

We have seen that wherever he went he had the faculty of inspiring respect and affection, and that an ever widening circle of friends admitted him to their intimacy, sought his advice, and confided in him with the perfect assurance of his ready sympathy.

A favorite Bible quotation of his was "Woe unto you when all men shall speak well of you." He deeply deplored the necessity of making enemies, but he early in his career became convinced that no man could accomplish anything of value in this world without running counter either to the opinions of honest men, who were as sincere as he, or to the self-seeking of the dishonest and the unscrupulous. Up to this time he had had mainly to deal with the former class, as in his successful efforts to establish the National Academy of Design on a firm footing; but in the future he was destined to make many and bitter enemies of both classes. In the controversies which ensued he always strove to be courteous and just, even when vigorously defending his rights or taking the offensive. That he sometimes erred in his judgment cannot be denied, but the errors were honest, and in many cases were kindled and fanned into a flame by the crafty malice of third parties for their own pecuniary advantage.

So now, having followed him in his career as an artist, which, discouraging and troubled as it may often have seemed to him, was as the calm which precedes the storm to the years of privation and heroic struggle which followed, I shall bring this first volume to a close.

END OF VOLUME I

Printed in the United States
By Bookmasters